THE TRUTH OF SCIENCE

ROGER G. NEWTON

THE TRUTH OF SCIENCE

Physical Theories and Reality

Viva Books

New Delhi | Mumbai | Chennai | Kolkata | Bangalore | Hyderabad | Kochi | Guwahati

First Indian Edition 2008
Reprinted 2010

VIVA BOOKS PRIVATE LIMITED

- 4737/23 Ansari Road, Daryaganj, New Delhi 110 002
  E-mail: vivadelhi@vivagroupindia.net, Tel. 42242200
- Plot No. 76, Service Industries, Shirvane, Sector 1, Nerul, Navi Mumbai 400 706
  E-mail: vivamumbai@vivagroupindia.net, Tel. 27721273, 27721274
- Jamals Fazal Chambers, 26 Greams Road, Chennai 600 006
  E-mail: vivachennai@vivagroupindia.net, Tel. 28290304, 28294241
- B-103, Jindal Towers, 21/1A/3 Darga Road, Kolkata 700 017
  E-mail: vivakolkata@vivagroupindia.net, Tel. 22836381, 22816713
- 7, Sovereign Park Apartments, 56-58, K. R. Road, Basavanagudi, Bangalore 560 004
  E-mail: vivabangalore@vivagroupindia.net, Tel. 26607409, 26607410
- 101-102 Mughal Marc Apartments, 3-4-637 to 641, Narayanguda, Hyderabad 500 029
  E-mail: vivahyderabad@vivagroupindia.net, Tel. 27564481, 27564482
- First Floor, Beevi Towers, SRM Road, Kaloor, Kochi 682 018, Kerala
  E-mail: vivakochi@vivagroupindia.net, Tel: 0484-2403055, 2403056
- 232, GNB Road, Beside UCO Bank, Silpukhuri, Guwahati 781 003
  E-mail: vivaguwahati@vivagroupindia.net, Tel: 0361-2666386

www.vivagroupindia.com

Published by arrangement with

Harvard University Press
79, Garden Street, Cambridge
Massachussetts 02138-1499
USA

ISBN 978-81-309-1381-0

Published by Vinod Vasishtha for Viva Books Private Limited, 4737/23, Ansari Road, Daryaganj, New Delhi-110 002. Printed and bound by Anand Sons, Delhi - 110 092.

TO RUTH

PREFACE

ALTHOUGH I was prompted to write this book by my irritation at the way a currently fashionable group of sociologists portray science and its results—a portrayal that has led to the so-called *science wars*—it is not intended as a polemic aimed at those who propagate such views; in only one chapter do I specifically go after them in detail. My purpose is constructive: to describe the intellectual structure of physical science and the understanding of reality that modern physics, the science with the most advanced and mature theoretical development, engenders. On a number of occasions, I will venture beyond that discipline, but some of the most important issues I tackle, particularly those concerning the role of theories and the nature of reality, assume a somewhat different cast in other sciences. As for the large question of truth, however, the outcome of the deliberations that I describe in terms of physics may well be taken to apply to all of science.

This book is intended for anyone with some scientific education; it is not addressed to professional philosophers or sociologists of science. No specific knowledge of physics on the part of the reader is assumed, and the many examples I draw on for illustrative purposes are fully explained. Some chapters will be more intellectually demanding than others, most particularly Chapter 10, which deals with the troublesome problem of reality at the submicroscopic level, where the quantum theory holds sway. I am afraid this is unavoidable; the questions at issue are difficult, even for physicists who routinely use quantum mechanics. There is no point in giving them a superficial presentation, with the mistaken implication that they are no different from other questions we will have to answer. Indeed, even though philosophers have struggled with many of the problems discussed for a long time, I may occasionally give the impression that their solution is simpler than it really is, and for this I apologize in advance to the reader. However, I see no reason for kicking up a lot of dust and then complaining one cannot see, as Leibniz accused many philosophers of doing.

I am indebted to many people for instructive discussions; among them I want to mention particularly Ciprian Foias, Howard Scott Gordon, Edward Grant, Noretta Koertge, and the late Richard S. Westfall. Most of all, I want to thank my wife Ruth for invaluable, tireless editing assistance in the writing of this book.

CONTENTS

THE TRUTH OF SCIENCE

INTRODUCTION

AS heirs of two clashing cultural progenitors, the Enlightenment (rational, orderly, measured) and the subsequent reaction of Romanticism (liberating, creative, irrational, and destructive), we approach the end of our millennium riding the wave of science but threatened by an undertow. From the crest we observe that the meteoric ascent of modern science allowed the West, and large parts of the rest of the world, to enjoy an economic prosperity previously unimaginable. The greatly accelerated pace of scientific advancement during this century has made all earlier human knowledge of the universe seem primitive and shallow. We have reason, now, to be confident that we understand a large part of the structure and constitution of the universe, from the interior of atoms to the farthest stars; we can successfully explain the mechanisms underlying the processes of matter and the forces between its constituents; and we are beginning to fathom the secrets of life from the gene to the brain.

The fruits of this knowledge in our "age of science" are visible everywhere; they have transformed our lives and, with the conquest of many dreaded and devastating diseases, doubled our life span. Communication through radio and television, transportation by auto and plane, transmission of information through computers have shrunk our planet into a global village. Justified or not, these developments have given scientific pronouncements an unprecedented authority; scientists are called upon to make judgments and predictions concerning the fears and hopes of a population that trusts them perhaps more than any other group in our society.

Nevertheless, we find everywhere, as well, a deplorable ignorance about the contents and character of science, which is identified more with technology and therapeutic medicine than with basic research: science is thought to be the improvement of television, the building of faster planes, the construction of more powerful weapons, or the curing of cancer and AIDS. And the ubiquitous confusion between science and its applications, between the plant and its fruits, leads some to imagine they can continue reaping the fruits while cutting back on the plant. Others see only the dark side of the rapid advance—increasingly destructive weapons, environmental degradation. A century ago there was a largely unalloyed enthusiasm for the harvest of technology (witness, for example, the public reaction to the World's Columbia Exposition in Chicago in 1893 and the St. Louis World's Fair in 1904). Today the reaction to science is just as likely to be hostility, as many in the West now question the value of science as the source and underpinning of a technological structure other parts of the world can only envy. The very success of science has spawned resentment against it.

This antagonism, aimed primarily at the physical and biological sciences, comes from two diametrically opposed directions. Those who are in despair over a widespread deterioration of moral and cultural values blame the skepticism and eternal uncertainty of science for eroding the comfortable feeling of certitude and security they drew from their spiritual beliefs. These critics envisage a return to a simpler age, in which people of faith, undistracted by an understanding of the world acquired through science, would take preachings based on religious authority as their sole guides to ethical and moral behavior. From the opposite direction, some practitioners of the more recent and less developed social and political sciences question the claim that the world truly is understood through science. They maintain that all of us, including scientists who in the past have been portrayed heroically as disembodied intellects, are creatures of our milieu: the origins of our ideas and the formulations of our ratiocinations and observations bear the imprints of ethnicity, gender, and class. Intellectual and philosophical arguments that start from valid observations, however, are often stretched to a point where they end up distorting—sometimes beyond recognition—what they are designed to illuminate, and so they are in this case. Influential sociologists announce with great confidence that the results of science—painstakingly gained by much observation, experimentation, and thought over the last four hundred years—have nothing to do with Nature and the external world under investigation, but are simply narratives, like myths and fairy tales, or the outcome of social agreements. Scientific "truths," they say, express the special perspective of the group from which they originate and are designed to further the group's political advantage.

The large majority of physical and biological scientists, of course, continue their work unperturbed. Nevertheless, in an age when pseudo-science flourishes, astrology parades as a science in the popular game Trivial Pursuit and shapes the daily schedule of a recent President of the United States, and some schools are forced to teach "creation science" as an alternative to the theory of evolution, the relativistic notions now fashionable at universities and among intellectuals exert a powerful influence on future legislators and the educated public. These ideas, hardly the harmless errors of a few misguided ignoramuses, are bound to have results detrimental to our society and corrosive to civilization as a whole. A world full of ignorance and superstition is a world full of fear, hatred, and panic.

That social influences exist—on the questions science asks and the problems it posits—can hardly be denied; this idea is neither new nor particularly controversial. I well remember Philipp Frank, a member of the Vienna Circle of positivists, talking about such ideas in his lectures at Harvard when I was a student almost fifty years ago. The results of science are not based on pristine apperceptions of naked facts, obtained by pure intellects working in isolated laboratories or ivory towers, but neither are they agreed-upon narratives or myths for political ends, linguistic artifacts produced in response to internal or external social pressures, as they are portrayed by some influential and vocal intellectual commentators today. Science stands or falls on the validity rather than the origins of its large structure of ideas. Those who, in light of the turbulent social currents in which we are all immersed, claim that the content of these ideas is of little rational relevance can be fairly accused of engaging in what the philosopher Larry Laudan calls "the most prominent and pernicious . . . anti-intellectualism in our time."[1] Operating most successfully at universities, they are robbing rational thought of all intellectual and cognitive value, leaving its expression a hollow rhetorical shell. This, ultimately, is why scientists, who value rationality above all else, are deeply offended by a misrepresentation that claims their work has as much epistemic value as the invention of fairy tales.

While arguing against this portrayal of science as myth, we should not assume that the scientific method that has evolved and flourished over the last 400 years was an inevitable development—it was valued consistently, after all, in only one culture. For this reason, we may regard it as a convention, albeit one with far-reaching consequences. My starting point is therefore the philosophical notion of *conventionalism.* In the first chapter I discuss this point of view within science itself—the notion that theories, and perhaps even experimental results, are largely conventions. In examining particular theories, I find that some parts, but certainly not all, are indeed conventional, though the assertion that even logic and mathematics are conventions proves unconvincing.

Chapter 2 takes up a recent version of conventionalism. I first examine the contention that science is subject to influences from sources other than Nature itself. It is certainly the case that extraneous social and political influences exist, though sociologists and historians may at times exaggerate their importance. But the malignant variant called "relativistic social constructivism" maintains, in its extreme form, that all scientific theories, and even their underlying facts, are social constructions quite uncorrelated with anything in Nature. I examine and criticize in detail the writings of some of the most prominent of these constructivists, with whose contentions I strongly disagree. The remainder of the book is devoted to the task of laying bare the structure and aims of the physical sciences, leading up to the question of their truth.

The primary aim of most physical scientists is to understand and explain the workings of Nature. In Chapter 3, I explore what is meant by an explanation, and I examine a variety of theories used for this purpose. Some of them, like the arrow of time in statistical thermodynamics, which keeps Humpty-Dumpty in bits and pieces, lead to the discovery of emergent properties, properties that were absent at a lower level. Most subfields of physics are formed by local theories derived from general ones on the basis of special approximations and assumptions. Since these theories are much closer to the phenomena, they lead their specialists to develop that most important quality for any scientist or mathematician, intuition. Unavoidably, there are hierarchies, not of value but of dependency, among the subfields and the various sciences; in that sense, all science is reductionist, and properly so.

In the fourth chapter I discuss several of the main tools scientists use for understanding. Models, which are taken seriously as descriptive of reality in some cases and less so in others, have always played an important part in scientific explanations. Analogies and metaphors are also important explanatory instruments that help us to understand novel phenomena in terms of familiar ones. I briefly examine a special category of theories that deal with history: geogony, cosmogony, and biological evolution. In my review of the anthropic principle, I find it wanting as a mode of scientific explanation of the universe.

By common assumption, science is based on facts. In the fifth chapter, I distinguish between individual and general facts, arguing that, outside the three history-related theories, only general facts are of interest to science. How are facts established? Many of them depend on theories, and therefore are said to be "theory laden," yet there are good reasons why they are, nevertheless, reliable and stable. I offer a number of examples of notorious pseudo-facts.

Chapter 6 explores the question of how theories arise from facts, emphasizing the distinction between the origins of theories and their establishment on the basis of evidence. Though the source of these theories lies in the imagination, which is often irrational and subject to social and psychological influences, their origin has no bearing on what earns them acceptance. The same can be said for mathematical theorems, whose proofs are independent of their imaginative source. How are theories tested? I examine in what sense Karl Popper's criterion of falsifiability may be regarded as more important than verifiability. Are there crucial experiments? What happens when theories are superseded? A useful general guide is provided by the "scientific method," but this method must not be interpreted as constraining. The most important criterion for or against acceptability in science is openly accessible evidence; for mathematics, acceptance is gained through general proofs that can be checked by every mathematician. This is what ensures the objective validity of the results and the stability of the ensuing structure.

Mathematics plays an enormously important role in science, most prominently in physics. From what source does its power derive, I ask in Chapter 7, and why is it so effective? Reviewing its changing historical role, we find that the computer now plays a significant auxiliary part in influencing what kinds of problems can be solved. The nature of mathematics determines its relation to science. Are theorems invented or discovered? What is a mathematical proof, and could modern science have developed using a kind of mathematics in which the idea of a proof is missing?

Causality, the primary explanatory principle of science, is the subject of Chapter 8, which begins with the remnant of Aristotle's notion of efficient causes left over after David Hume had ripped it apart. As a matter of universal experience, causes always precede their effects, a characteristic that is crucial in some areas of physics. I review the doctrine of determinism and its modern origin in Newton's equations, concluding that the definition of the state of a physical system subject to determinism depends on the nature of these equations. An analysis of the quantum theory shows that, despite frequent assertions to the contrary, it, too, is deterministic. Knowing the quantum state of a system, however, is different from knowing a classical state. At this point probablilites enter the picture. I discuss the notion of probability in some detail, including the frequency definition and Popper's propensity theory, both of which have an important bearing on the interpretation of quantum mechanics and the view of reality to which it leads.

The last three chapters explore more deeply the basic problems arising in the physical sciences concerning reality and truth. After contrasting realism with idealism, Chapter 9 turns to the entities that are considered real in classical physics. Though their reality remained unclear in the minds of some scientists, two concepts were dominant in the nineteenth century: particles, which, as Democritus had taught long ago, formed the basis for the structure of all matter, and fields, introduced by Michael Faraday, which transmitted the forces between the particles. Doubts about the reality of all fundamental physical notions began to intensify early in the twentieth century, when Einstein's theory of relativity raised questions concerning "real" length and time, but it was the quantum theory—its particles without identity or trajectories of motion—that brought forward the most basic problems about reality. Everything now seems dissolved by the universal wave-particle duality, and we wonder what is meant by the reality of minute objects that "live" for a tiny fraction of a second and then decay, or by the existence of particles such as quarks that can never be found in isolation. Much of what is real appears to become "virtual," and I conclude that realism depends on the scale of the beholder's view. The principal difficulty arises from the limitations of our language, which is tied to the scale of everyday life and seems ill-adapted to the micro world closed to our senses.

Chapter 10 delves into the difficult problems raised by physical reality at the submicroscopic scale, where we have no choice but to confront the puzzles and paradoxes of the quantum. After a discussion of the wave function, its interpretation and its mysterious "collapse," I turn to the most serious reality questions in the quantum theory. With Bohr and Einstein on opposing sides, the famous EPR debate and "entanglement" are introduced. Bell's inequality presents a way of subjecting to experimental test Einstein's search for the existence of "elements of reality" without "spooky action at a distance," and the evidence favors Bohr. Quantum field theory comes to the rescue by automatically producing both particles and waves, if we prefer using these concepts. My view of submicroscopic reality is based on this quantum field, while I conclude that both particles and waves are manifestations of our inadequate language.

Finally, Chapter 11 takes up the concept of truth as it applies to science. Distinguishing between the definition of truth and criteria for recognizing it, I adopt coherence as a test—a body of assertions is true if it forms a coherent whole and works both in the external world and in our minds. Science gradually approaches but never arrives at a truth that is, above all, public and openly sought. Attacks by cynics and ideologues notwithstanding, objectivity is an indispensable constituent goal, for which the scientist must strive in spite of personal biases, difficult as these often are to overcome. The pursuit of truth and the ideal of objectivity—purposes and values that scientists implicitly adopt and carry with them, not always with conspicuous success—constitute what might be called the "scientific attitude." Despite persistent current criticism, that attitude has served civilization well.

CONVENTIONS

EVIDENCE obtained by experimentation for all to see, and general proofs sturdy enough to withstand scutiny—requirements neither obvious nor congenial to other cultures—were the foundation stones on which the ancient Greeks grounded our understanding of Nature and our knowledge of mathematical relations. The kind of mathematics pursued by the Babylonians, the Egyptians, the Hindus, and the Chinese led to many insights but never contained the idea of a *proof* as we know it. And while all these civilizations developed important technological advances through trial and error—watching and testing rather than simply following tradition—they did not arrive at general propositions about Nature grounded on observation and experiments that could be replicated, analyzed, and argued over. Rather, their views of Nature depended more on sacred books, the authority of prophets, private experience, or pure thought alone. The physicist Alan Cromer argues in *Uncommon Sense: The Heretical Nature of Science* that this Greek methodology was antithetical to what he calls the "egocentric" manner of gathering knowledge that pervaded other cultures and still largely dominates most of humanity: "Scientific thinking didn't—and couldn't—evolve from the prophetic tradition of Judaism and Christianity; it arose from a totally different tradition."[1] Similarly, the biologist Lewis Wolpert is convinced that "it is almost universal among belief systems not influenced by the Greeks that man and nature are inextricably linked, and such philosophies provide a basis for human behaviour rather than explanations about the external world."[2]

This invaluable innovation of the Greek culture lay dormant for centuries, held in memory, translated, and preserved by the Arabs. Later it was reintroduced to the intellectual consciousness of Europe by translations of the Arab texts into Latin in the late Middle Ages, induced to grow during the Renaissance, and brought to full flower in modern science. "The development of Western science," Einstein wrote in a letter,

> has been based on two great achievements, the invention of the formal logical system (in Euclidean geometry) by the Greek philosophers, and the discovery of the possibility of finding out causal relationships by systematic experiment (at the Renaissance). In my opinion one need not be astonished that the Chinese sages did not make these steps. The astonishing thing is that these discoveries were made at all.[3]

Thus modern science, Einstein, Wolpert, Cromer, and others argue persuasively, is not a natural way of looking at the world, bound to emerge among civilized people, but a very special, enormously productive methodology that historically arose only once and was fortunate to survive a long and perilous dormancy. Its emergence was neither inevitable nor its value immediately obvious. In fact, from the beginning it was strongly resisted, and it is resisted to this day, not only by religious fundamentalists but also by fashionable political groups. Opposition comes, for example, from New Age adherents and from radical feminists, whose science projects, the philosopher Sandra Harding declares, "emphasize personal experience as a source of knowledge."[4] But personal experience that cannot be publicly replicated is precisely the kind of evidence that has no place in modern science.

The Scientific Method as a Convention

The dispute between Robert Boyle and Thomas Hobbes offers a good historical example of the controversial nature of the rise of the concept of experimental science in the seventeenth century. Boyle had perfected the construction of a pump to evacuate the air from a vessel, producing a much better vacuum in his laboratory than had previously been available and allowing him to perform experiments with gases at various pressures. Among other conclusions, his data led to what is now called *Boyle's law:* so long as the temperature is held fixed, when the pressure or volume of a gas is changed, the product of the two remains constant.

Boyle is generally credited not only with specific discoveries, however, but with the development of the whole notion of laboratory science, the idea that experiments were not simply demonstrations performed by well-dressed gentlemen in front of an audience for the purpose of persuasion but were procedures for generating answers to questions about Nature. When Boyle's findings conflicted with those of others (Christiaan Huygens, for example), he relied on the superior quality of his air pump to give him a dependable answer that could be verified by those witnessing the experiment. He thus "question[ed] not [Huygens's] Ratiocination, but only the stanchness of his pump."[5] Anyone who had a pump as good as his could repeat the experiment and would obtain the same result. This important new line of argumentation had the additional virtue of being less personal.

The novel procedure of answering "philosophical" questions by resorting to witnessed and repeatable experimental tests was strongly attacked by Thomas Hobbes, to whom the vacuum was a metaphysical concept. In his view, what Boyle was doing had no philosophical relevance—his methods were not only wrong, they were actually dangerous. Instead of depending upon rational thought, Boyle's experiments had to be done with an expertly constructed piece of apparatus and witnessed by members of the Royal Society. As Steven Shapin and Simon Schaffer put it in their study of this controversy, "Hobbes maintained that the experimental form of life could not produce effective assent: it was not *philosophy.*"[6] In Hobbes's way of thinking, only rational argument mattered, and empirical data were regarded as ephemeral: "Hobbesian philosophy did not seek the foundations of knowledge in witnessed and testified matters of fact: one did not ground philosophy in 'dreams'."[7] The clash between Boyle and Hobbes had, of course, been foreshadowed long ago by that between Aristotle and Plato.

What such controversies show is that the method of science as we apply it now does not force itself upon the human mind as either logically necessary or inevitable. Therefore it would be fair to call it a convention. Science demands that "its standardised procedures be adhered to," David Bloor writes. "These procedures declare that experience is admissable only in as far as it is repeatable, public and impersonal. That it is possible to locate experience that has this character is undeniable. That knowledge should be crucially linked to this facet of our experience is, however, a social norm . . . Other activities and other forms of knowledge have other norms."[8] Indeed, in many cultures, both old and contemporary, knowledge is not assumed to be based on scientific procedures.

To say that something is a convention, however, as many conventionalists have stressed, is not necessarily to imply that it is a *mere* convention. By agreeing that the adoption of the scientific method is a convention, I do not mean to say that it is inconsequential or completely arbitrary, only that some people and some cultures have not adopted it and do not wish to put it to use in acquiring knowledge. That choice, however, has far-reaching intellectual and practical consequences. On the one hand, it has led both to a vast enrichment of our understanding of Nature and to all the benefits that flow from technology and therapeutic medicine based on science; on the other hand, it has led to what some regard as spiritual impoverishment and to the deleterious side effects of technology.

Conventionalism within Science

If we can agree that the adoption of the scientific method is a convention, must we conclude that the results obtained by this method—the laws and theories of science—are also conventions? This is the fundamental question raised by the school of conventionalism, called *nominalism* in its extreme form, which has sprouted malignant variants among some influential contemporary thinkers. All scientific results and theories are conventions, they contend, with the implication, at least in the minds of some, either that these results say nothing about the real world at all or that Nature and reality are simply *defined* by these conventions.

Einstein, on several occasions, expressed sentiments that superficially appear to be conventionalist: "Science," he wrote, "is . . . a creation of the human mind, with its freely invented ideas and concepts";[9] theories, he said in his 1933 Spencer lecture, are "free inventions of the human intellect." This phrase, however, it is important to note, appeared in a context that limited its validity:

> The structure of the [theoretical system of physics] is the work of reason; the empirical contents and their mutual relations must find their representation in the conclusions of the theory. In the possibility of such a representation lie the sole value and justification of the whole system, and especially of the concepts and fundamental principles which underlie it. Apart from that, these latter are free inventions of the human intellect, which cannot be justified either by the nature of that intellect or in any other fashion *a priori*.[10]

Furthermore, it appears that what seems like conventionalism is in reality an expression of a Leibnizian philosophy:

> No one who has deeply thought about this subject will deny that the world of sense perceptions practically determines the theoretical system uniquely, even though there is no logical path leading from the perceptions to the fundamental laws of the theory; this is what Leibniz so beautifully called "pre-established harmony."[11]

It would therefore be a mistake to regard Einstein's view of science as conventionalist.

The great French mathematical physicist Henri Poincaré, on the other hand, was indeed a conventionalist, and his arguments are worth quoting at some length:

> We shall also see that there are several sorts of hypotheses; that some are verifiable, and once confirmed by experiment become fruitful truths; . . . that others, finally, are hypotheses only in appearance and are reducible to disguised definitions or conventions.
>
> These last are met with above all in mathematics and the related sciences. Thence precisely it is that these sciences get their rigor; these conventions are the work of the free activity of our mind, which, in this domain, recognizes no obstacle. Here our mind can affirm, since it decrees; but let us understand that while these decrees are imposed upon *our* science, which, without them, would be impossible, they are not imposed upon nature. Are they then arbitrary? No, else were they sterile . . .[12]

More specifically, Poincaré contends that

> The principles of mechanics, then, present themselves to us under two different aspects. On the one hand, they are truths founded on experiment and approximately verified . . . On the other hand, they

> are postulates applicable to the totality of the universe and regarded as rigorously true.
>
> If these postulates possess a generality and a certainty which are lacking to the experimental verities whence they are drawn, this is because they reduce in the last analysis to a mere convention . . .
>
> This convention, however, is not absolutely arbitrary; it does not spring from our caprice; we adopt it because certain experiments have shown us that it would be convenient.[13]

Poincarè also, however, recognizes that there are limits to the extent of conventionalism:

> Some people have been struck by this character of free convention recognizable in certain fundamental principles of the sciences. They have wished to generalize beyond measure, and, at the same time, they have forgotten that liberty is not licence. Thus they have reached what is called *nominalism,* and have asked themselves if the savant is not the dupe of his own definitions and if the world he thinks he discovers is not simply created by his own caprice. Under these conditions science would be certain, but deprived of significance.
>
> If this were so, science would be powerless. Now, every day we see it work under our very eyes. That could not be if it taught us nothing of reality.[14]

To understand both the validity and the limitation of these arguments for the proposition that scientific laws contain an element of convention, it is instructive to look at a prototype—Newton's laws of motion—which will show in detail to what extent it may be regarded as conventional. It might be a good thing to refresh our memories by restating the three laws:

(1) When no forces act on an object, it remains at rest or in a state of uniform motion. (2) Acceleration—changing the state of rest or uniform motion of an object—requires the action of a force of a magnitude proportional to and in the same direction as the acceleration, the constant of proportionality being the object's inertial mass; in mathematical form, the law is written as $F = ma$. (3) To every applied force there is an equal and opposite force of reaction.

All three of the quantities appearing in the second law—force, mass, and acceleration—have conventional aspects. One may well argue that this law *defines* what we mean by a force, an argument that becomes particularly apparent from the perspective of the special theory of relativity, where the concept of force is not *a priori* clear. There, the second and third law are used as the *definition* of a *Newtonian force.* In the Newtonian view, a force is assumed, or defined, to be present whenever the velocity changes: the first law states that if there is no force, the velocity of the object remains constant, but not necessarily zero, contrary to Aristotelian ideas and to most of our everyday experience. It would be an overstatement, however, to draw from this line of reasoning the conclusion that Newton's laws of motion are nothing but a way of defining the concept of force. After all, everyone has had some experience with forces, and the notion of a force as a push or a pull has a definite intuitive meaning. We feel the force of a weight, the pull of a string, or the effort required to stretch a spring. Furthermore, the concept of force is also used in statics, where its magnitude can be measured independently of the laws of motion. Therefore, the left-hand side of the equation expressing the second law is not a concept that is newly coined and defined; it is also supported by an anthropomorphic idea and by an independent use elsewhere.

The notion of *acceleration,* on the other hand, has little intuitive meaning, as any physics teacher who has tried to make students understand the difference between acceleration and velocity will know. It turned out to be enormously important for the mathematical development of the laws of motion that Newton defined acceleration in a way already customary during the Middle Ages, as the rate of change of the velocity *with respect to the time* rather than with respect to the distance. Since Galileo had come to the conclusion that the acceleration, *so defined,* of a falling object was constant, any alternative definition would have very much complicated matters.

Finally, there is the inertial *mass* of the object. This is the one physical quantity that may truly be said to be defined by the second law of motion. The principal content of the law, of course, is the proportionality of the force and the acceleration (and that they point in the same direction). The use of the same word *mass* for the constant of proportionality and for the property of matter that makes it subject to the gravitational attraction has led to all sorts of confusion, but the problem is at least in part built into our language: we call an object that is hard to move *ponderous*—a word whose origin means *heavy*—

that is, "subject to a strong gravitational force." (The very fact, which Newton already found remarkable, that the property of resisting acceleration and the property of being subject to gravitational attraction can be used interchangeably became one of the cornerstones of Einstein's general theory of relativity 250 years later.)

So much for the conventions involved in Newton's laws of motion. Would it then be fair to say that the laws are *only a convention* and tell us nothing about the world? Of course not. That the definitions contained in and implied by them are enormously fruitful and convenient has vast implications for the structure of Nature. At the same time, the conventional elements testify to the fact that Newton's laws of motion are not simply inductive consequences of observations but are products of a very fertile imagination. From this I conclude that Poincaré was right, both in saying that scientific laws contain conventional elements and in denying that, as a consequence, all the results of science are conventional.

The principal opportunity for conventions to enter into our conceptions of the world is the fact that all theories are based on a *simplification* of Nature: no theory fits Nature with perfect accuracy. The trajectory of a real baseball, for example, is not described precisely by simply solving Newton's equation of motion; describing it realistically, including all the influences of air resistance, air turbulence, wind pressures, changes in the gravitational force due to inhomogeneities in the ground and to altitude, etc., is enormously complicated. The Newtonian laws apply directly only to the motion of an idealized baseball, moving in uniform gravity through a vacuum and with no disturbing influences; the other complications are taken into account by subjecting the idealized motion to "corrections," which, however, are again based on these laws. Newton's formulation derives from an idealized world that is very much simpler than reality. Such simplifications, which other sciences have not yet and may never be able to duplicate, make physics the powerful tool it is. But clearly, simplification requires omission, and the selection of what remains contains an element of convention. The Aristotelian view that a force was necessary to sustain motion proves to have been another convention, closer to everyday experience but much less productive; the corrections that would be necessary to calculate a ball's actual trajectory, if we started with an Aristotelian idealization, would be very large and much harder to apply. Moreover, each of the corrections applied in the Newtonian case—air resistance, air turbulence, wind pressures, and so on—we can ascribe to physical causes, which would be very difficult to do with an Aristotelian description.

The Conventionalist Stratagem

There is, however, an aspect of conventionalism that has more disturbing consequences. What happens when reliable experimental observations seem to contradict an established theory? To the question what would be the fate of his general theory of relativity if the predicted gravitational red shift in massive stars were not observed, Einstein is said to have replied "The theory will be dust and ashes."[15] But in reponse to the finding that the theory implied an expansion of the universe, for which there was no evidence at the time, he introduced the "cosmological constant," an *ad hoc* modification which he later regretted and considered to be ugly.

Adopting an attitude that the laws of physics are, after all, only conventions makes it easy to evade conflicts between theory and experiment by "tinkering" with the theory. Karl Popper recognized this as the principal danger of conventionalism, which he regarded as impossible to refute. However, since such tinkering gets us nowhere, his advice is simply to reject it. "The only way to avoid conventionalism is by taking a *decision:* the decision not to apply its methods. We decide that, in the case of a threat to our system, we will not save it by any kind of *conventionalist stratagem.*"[16]

The discovery by T. D. Lee and C. N. Yang of the breakdown of the law of parity conservation some forty years ago was an excellent illustration of the avoidance of the conventionalist stratagem. Parity is the property of a system, such as a quantum-mechanical particle, that represents its response to the operation of reflection in a mirror. The parity "quantum number" of a particle is either +1 or −1. It was a universal assumption of long standing in physics that the laws of nature were "of course" invariant under reflection, that is, the mirror image of any event allowed or predicted by the laws should also be allowed and predicted to occur with the same probability. This assumption leads to "parity conservation" as a mathematical consequence: the parity of any system is the same before and after an event such as the decay of a particle. Experimental observations of the decay of two particles, then named *tau* and *theta*, led to what was called the "tau-theta puzzle": the two particles seemed to have identical properties, such as mass and spin, but their decay products possessed opposite parities. Therefore, on the assumption of parity conservation, the tau and the theta had to have opposite parities too—they could not be the same particle. It was regarded as extremely strange that Nature would produce two different particles with exactly the same properties, except for their parities. This Gordian knot was cut by Lee and Yang with the suggestion that the tau and the theta were, in fact, one particle and its decay did *not* conserve parity.

Needless to say, this thought had previously crossed the minds of others, but it had always been dismissed as an *ad hoc* modification; such a solution to the puzzle seemed to be exactly the "conventionalist stratagem" of which Popper had warned and which good scientists try to avoid at all costs. Introducing an *ad hoc* assumption to "save the phenomena" does not explain anything and is therefore scientifically useless, *unless it leads to other consequences that can be experimentally checked.*

And that is exactly what Lee and Yang, to their great credit, proceeded to do. The force of interaction regarded as responsible for the decay of these particles was the same "weak force" that is also responsible for nuclear radioactivity, called beta decay. Lee and Yang therefore searched the literature and found that although conservation of parity had always been taken for granted in nuclear beta decay, among the many experiments in this area of physics no specific test of that question had ever been performed.[17] So, in the face of skepticism and even derision by many prominent physicists, they suggested a specific experiment that would subject the question of parity conservation in beta decay to a clear test, an experiment soon performed by C. S. Wu and her collaborators, and indeed, parity conservation was found to be badly violated—the world did not look the same in a mirror. The following year, Lee and Yang received the Nobel prize, and everyone else ate crow.

Conventionalism of Logic and Mathematics

If there is a grain of truth in the contention that scientific theories contain elements that are conventional, I don't think the same can be said for the stronger assertion that even logic is, at least in part, a matter of convention. "Just as men haggle over questions of duty and legality," writes David Bloor, "so they haggle over questions of logical compulsion."[18] To bolster his claim of the conventional nature of logic, Bloor cites the description by the anthropologist E. E. Evans-Pritchard of the Azande society in central Africa and the reasoning that is said to prevail there.

The Azande consult oracles for all questions regarded as important and ascribe all unfortunate events to the evil influences of witches. In order to decide whether an event is caused by witchcraft, they consult an oracle in the form of a poisoned chicken: yes if the chicken lives, no if it dies. They also believe that being a witch is an inherited property of a person, residing in a witchcraft substance in the belly, and is transmitted by males to all their sons, by females to all their daughters. With the help of oracles, autopsies reveal the presence or absence of the evil stuff in cases of persons accused of being witches.

Now here is the problem: it would seem once someone has been found to be a witch, the Azande ought to conclude that a whole line of this person's descendants are also witches. Similarly, if an autopsy of a man failed to reveal the presence of witchcraft substance, the evidence should absolve his entire clan. As a matter of fact, the Azande draw that conclusion only for close kinsmen. Writes Evans-Pritchard:

> To our minds it appears evident that if a man is proven a witch, the whole of his clan are ipso-facto witches, since the Zande clan is a group of persons related biologically to one another through the male line. Azande see the sense of this argument but they do not accept its conclusions, and it would involve the whole notion of witchcraft in contradiction were they to do so. In practice they regard only close paternal kinsmen of a known witch as witches. It is only in theory that they extend the imputation to all of a witch's clansmen.[19]

Discussing the Azande logic, Bloor refers to Wittgenstein's equating of the drawing of logical conclusions with "thinking that something cannot be otherwise"; since the Azande regard the proposition "that the whole of a witch's clan cannot be witches" as something that cannot be otherwise, they are logical, he contends, when they do not draw from the heritability of the witchcraft substance the conclusion which we regard as logical, namely that the entire clan must be witches. Bloor therefore declares "there must be more than one logic: an Azande logic and a Western logic."[20] In my view, this argument is unconvincing and the conclusion unjustified. The Azande, like the large majority of people all over the world, simply do not put a high value on logical consistency. That, indeed, was essentially how Evans-Pritchard explained it: "they have no theoretical interest in the subject."[21] Anyone listening to the pronouncements of our politicians should surely be familiar with such behavior. Is there, therefore, a separate "politicians' logic"?

The multivalued logic introduced by the mathematician E. L. Post[22] furnishes another example in which logic is taken to be conventional: it contains not only the two truth-values *true* and *false,* but *m* gradations of them. Hans Reichenbach[23] applied the special case of a three-valued logic to his interpretation of the quantum theory in trying to cope with the fact that certain statements entering physics that are perfectly meaningful from a classical point of view are not regarded as allowed in quantum mechanics. An example of such a statement might be "This particle now has precisely position x and momentum p." Heisenberg's indeterminacy principle rules out such statements—the product of the precisions with which position and momentum can be specified simultaneously has to be greater than Planck's constant[24]—and they are therefore generally considered to be without meaning. Reichenbach's scheme, rather than regarding them as meaningless, assigns them the truth-value "indeterminate," that is, neither true nor false, analogously to the verdict "not proven" in Scottish law. This interpretation of the quantum theory did not really catch on. Any such construction of a distinct logic for a special purpose seems to me quite artificial and cannot be regarded as any indication that the two-valued logic we normally employ is merely a convention. One also has to note that the reasoning and the proofs of theorems in multivalued logic employ the conventional two-valued variety.

Within mathematics, the intuitionists, to be discussed further in Chapter 7, effectively employed in their argumentation a system of three-valued logic. One of the powerful devices sometimes used to prove a theorem is the *reductio ad absurdum*—assuming the theorem to be false leads to a contradiction: therefore we have no choice but to conclude the theorem is correct. The intuitionists deny that we have no choice and regard such proofs as invalid.

There are sociologists of science who go further and regard all mathematical proofs as conventions. As an example Bloor[25] offers the well-known proof by Aristotle that the square root of two cannot be a rational number (expressible as a quotient of two integers). Assume that $\sqrt{2} = p/q$, where p and q are incommensurate integers (that is, they have no common divisor and the ratio has been reduced to its simplest form). It follows that $2 = p^2/q^2$ and, hence, $p^2 = 2q^2$. Therefore p^2 is an even number, and since the product of two odd numbers is odd, we conclude that p must be even: $p = 2a$, where a is some other integer. But then $p^2 = 4a^2$, and we find by cancellation that $2a^2 = q^2$. This equation, however, implies by the same reasoning used above that q, like p, must be an even number, which contradicts the original asssumption that p and q have no common divisors. It follows therefore that the original assumption must be incorrect, and $\sqrt{2}$ cannot be expressed as the ratio of two whole numbers.

To the Greeks, Bloor points out, Aristotle's argument (which, incidentally, relies on the technique of *reductio ad absurdum*) proved that $\sqrt{2}$ is not a number, while to us it proves that it is an irrational number. This is, of course, correct, but the successive enlargements of the number system, which grew from the natural numbers to include the rationals, then irrationals, then negatives, then transcendental numbers, and finally imaginary numbers, made up a chain of creative steps that led to an enormous enrichment of mathematics. The system of classifying numbers may be called conventional because it was creative, but there is no evidence whatever that it could just as well have been done differently. Each of these steps might not have been taken, but then large parts of mathematics would never have come into being. In fact, a Platonist would surely call them *discoveries* of new kinds of numbers, and they would lose their conventional character altogether.

Bloor makes a further point that a mathematics which "never set much store in the categories of odd and even" might even conclude from Aristotle's proof that numbers could be both odd and even! This, of course, is absurd: whether you "set much store in the categories of odd and even" or not, it is easily proved that a number cannot be both odd and even, that is, both divisible by two and not divisible by two. One indispensable requirement of a mathematical system is consistency.

Bloor discusses another example—a famous theorem by Leonhard Euler—which is instructive in its own right because it illuminates the nature of a mathematical proof, though not in the way Bloor intends. For any polyhedron, that is, any finite solid figure bordered by plane surfaces, the numbers of vertices (corners) V, edges E, and faces F are related by the simple equation $V - E + F = 2$. Euler convinced himself of the correctness of this formula and stated it as a theorem on the basis of examining a large number of instances and finding it to be true in each case. Such reasoning is not regarded as a real proof, because there might, after all, be a contrary instance that was overlooked, and some sixty years later Augustin-Louis Cauchy announced an ingenious argument that required replacing the surface of the polyhedron by a rubber sheet, removing one face, and flattening it out. Not long after Cauchy's demonstration, two mathematicians found examples of polyhedra that eluded his proof. Did this mean that Euler's theorem was incorrect? One could interpret these counterexamples in such a way and let it go at that. Instead, however, a somewhat more restrictive definition of what constituted a polyhedron was introduced, to which the counterexamples did not apply. New counterexamples led to further restrictions, and this process went on for decades. Bloor sees this as demonstrating that mathematical proof, far from being ironclad and tight, is a matter of social convention, subject to negotiation, but this unjustified conclusion reveals that he does not understand one of the chief functions of a mathematical proof: to tightly circumscribe the area of a theorem's applicability.

The standards mathematicians use for accepting a proof as airtight and rigorous have evolved in the course of time, and during the nineteenth century they were raised to a point where they have become part of the friction between the disciplines of pure and applied mathematics. The rigorous demands of pure mathematics now require, in principle, that every proof could be replaced by a chain of elementary logical steps leading from the hypotheses to the conclusion. It is this completely explicit statement of all the hypotheses entering a rigorous mathematical proof, and the avoidance of tacit and unconscious assumptions, that Poincaré had in mind when he emphasized the usefulness of mathematics to science. Does this historical development of the standards of proof imply that proofs are a matter of convention? Not unless we regard any improvement or refinement as a convention. There can be no question that the prevailing norms are *higher* than the old ones, and not just different. Since future mathematicians would never agree to return to more relaxed standards, in the sense of accepting a proof of a theorem that is known to be false under more rigorous tests, it is quite wrong to call mathematical proofs conventions.

Many theorems, including Euler's, had been accepted on the basis of earlier, less rigorous proofs. In none of these cases were the theorems simply discarded when counterexamples were found. All that needed to be done was to state the hypotheses more explicitly and thus delimit the class of objects to which a given theorem applied. Such limitations, in turn, often gave rise to productive new concepts as well as distinctions between classes of mathematical entities that had previously been unknown, thereby stimulating progress.

There is no plausible argument or credible evidence for a far-reaching claim that either logic or mathematics is a matter of convention. The only general aspect of mathematics that may perhaps be regarded as conventional is the use of *proofs*. Just as the adoption of the methods of modern science for accepting new knowledge may be called a convention, so may the requirement of strict proofs in mathematics. Neither of them is adopted by all civilizations, and both have powerful consequences. The nature of these proofs and the reasoning employed in them, however, are conventional only to a very limited extent. To accept methods of proof ruled out of bounds by the intuitionists is a convention, and an extremely beneficial one. But there is no reason to expect that future contact with alien scientists and mathematicians might find them using a different logic and reasoning, accepting theorems which we regard as false or claiming some of our theorems to be incorrect.

Let me now turn to an outgrowth of conventionalism that has become fashionable during the last twenty years, not among those practicing the physical or biological sciences but among sociologists studying and commenting upon them.

2

SCIENCE AS A SOCIAL CONSTRUCT?

SINCE the adoption of the scientific method and the formulation of its results are to a certain extent arbitrary, and since both sometimes play enormously important roles in shaping society's view of the world, it can come as no great surprise that they would be subject to criticism and assault on political grounds. Those groups and forces in society that rely on a *Weltanschauung* with which the procedures of science or its conclusions are in conflict will attempt to discredit scientists intellectually or silence them altogether. That, of course, has happened before in the course of history—Galileo Galilei was tried and convicted in the seventeenth century because his scientific views were considered dangerous to the established beliefs of the Church.[1] As Shapin and Schaffer argue, the conflict between Hobbes and Boyle, played out on an intellectual level, also contained a strong political element determined by the struggles during the Restoration of the English monarchy.

> [T]he disputes between Boyle and Hobbes became an issue of the security of certain social boundaries and the interests they expressed. For Boyle this would inevitably involve the connection between the work of the experimental philosopher and that of the Priest as Christian apologist. Their functions reinforced each other and Hobbes was their common enemy. But for Hobbes any profession that claimed such a segregated area of competence, whether priestly, legal, or natural philosophical, was thereby subverting the authority of the undivided state. The events of the Restoration made that authority a vital concern for philosophy.[2]

But it should not be concluded that these kinds of conflicts and attitudes are things of the past and have no modern counterparts. We can readily find more recent examples, beginning with the political and religious attacks on Darwin's theory of evolution, which are continuing today.

More specifically relevant are the vicious attacks leveled at Einstein and his theory of relativity in Germany in the 1920s, which became state policy under the Hitler government. This theory (whose name was misunderstood by many to imply a general moral relativism) was regarded as an outgrowth of the "Jewish mind," a charge made even by some prominent German physicists, one of whom, the Nobel Laureate Philip Lenard, published a book entitled *German Physics.* The Nazis, suspicious of science in any case, regarded the race or ethnicity of scientists as a strongly determining factor in the theories they developed,[3] and keeping science "Aryan" was considered politically important.

The view taken in the Soviet Union, where the situation was quite different from the one in Nazi Germany, offers another instance of political contamination. There, the *a priori* assumption was very much in favor of science, since Marxism, after all, was regarded as a "scientific" view of the world. Since the philosophy underlying Marxism was "materialistic," though, anything that deviated, or was thought to deviate, from "materialism" was suppressed. When the interpretation of quantum mechanics became a controversial topic—and Bohr's "Copenhagen interpretation" was regarded by many as tinged with the philosophy of "idealism," anathema to materialists—it was not allowed to be published and taught in the Soviet Union. Some textbooks of physics, whose earlier editions had violated this taboo, were changed in later editions, with fulsome praise for the wisdom of Lenin and Stalin in their prefaces. Soviet chemists also rejected some of Linus Pauling's ideas on ideological grounds, but, because of the Marxist predilection in favor of Lamarckian evolution (enhanced by the personal power of Lysenko), the biological sciences, and especially genetics, were subject to the most severe political pressures.

In this country and in other Western democracies, of course, we have not had any government-sanctioned attacks on science or scientists *qua* scientists, but politically motivated attacks have been mounted by groups of people with a variety of ideologies. For example, extreme feminists claim that much of what scientists say is determined simply by the fact that men dominate science. This is sometimes referred to as the "perspectivist" account: "women (or feminists, whether men or women) *as a group* are more likely to produce unbiased and objective results than are men (or nonfeminists) as a group,"[4] claims Sandra Harding, the author of the aphorism "Science is politics by other means." Attacks come, as well, from those unhappy with "scientific liberalism." The journalist Bryan Appleyard sees himself as part of a "struggle to hold back the cruel pessimism of science," a struggle which is "a long tale of decline and defeat."[5] To him, scientists "who insist that they are telling us how the world incontrovertibly is are asking for faith in their subjective certainty of their own objectivity."[6]

The Sociological Perspective

The emergence of modern science, I argued in the last chapter, was not an inevitable development and it did not occur everywhere. It therefore becomes an intriguing sociological problem, studied prominently by Robert Merton, to account for its appearance in our culture. This, however, is not the topic of interest to many contemporary social scientists, who prefer to focus their attention on the *contents* of science. Historians and sociologists have long written and speculated about the role societal, personal, and other external conditions have played in the development of scientific ideas. Darwin's theory of evolution, for example, is known to have been influenced to some extent by the ideas of the political economist Thomas Robert Malthus.[7] Scientists do not grow up in Skinner boxes, nor do they hatch their ideas in splendid isolation. Furthermore, social surroundings and cultural conditions surely have something to do with the kinds of questions they ask, the judgments they make about the relative importance of problems to be addressed, and the metaphors they use when presenting a theory or an explanation.[8]

Who could deny that the invention of the steam engine and its consequences in Europe and the United States—the industrial revolution—molded the formative years of the science of thermodynamics and shaped the language in which it has been couched from its beginning? Thermodynamic terminology is permeated with words like "heat reservoir," "Carnot engine," "perpetual motion machine," and the like. The main works by two of the principal contributors to the newly emerging science of heat in the early nineteenth century—Sadi Carnot, a French engineer who did not just want to apply the effects of heat to generate industrial power but to understand it, and the German physicist Rudolf Clausius—open with references to the steam engine. Many of the metaphors used in thermodynamics refer to industrial paraphernalia, engines, machines, and the efficiency of their operation, which is one of the reasons why Percy Bridgman thought these laws "smell more of their human origin"[9] than other laws of physics.

Nevertheless, the influence of metaphors in science has been exaggerated beyond bounds. Harding, quoting Francis Bacon, charges early science enthusiasts with rape and torture imagery: "For you have but to follow and as it were hound nature in her wanderings, and you will when you like to lead and drive her afterward to the same place again . . . Neither ought a man to make scruple of entering and penetrating into those holes and corners when the inquisition of truth is the whole subject."[10] Therefore, she argues,

> if we are to believe that mechanistic metaphors were a fundamental component of the explanations the new science provided, why should we believe the gender metaphors were not? A consistent analysis would lead to the conclusion that understanding nature as a woman indifferent to or even welcoming rape was equally fundamental to the interpretations of these new conceptions of nature and inquiry. Presumably these metaphors, too, had fruitful pragmatic, methodological, and metaphysical consequences for science. In that case, why is it not as illuminating and honest to refer to Newton's laws as "Newton's rape manual" as it is to call them "Newton's mechanics"?[11]

Needless to say, the effects of these insidious metaphors that she sees in Newtonian mechanics are devastating to women:

> Both nature and inquiry appear conceptualized in ways modeled on rape and torture—on men's most violent and misogynous relationships to women—and this modeling is advanced as a reason to value science . . . As nature came to seem more like a machine, did not machines come to seem more natural? As nature came to seem more like a woman whom it is appropriate to rape and torture

> than like a nurturing mother, did rape and torture not seem a more natural relation of men to women?[12]

Can anyone who knows anything about physics possibly recognize that science in these tormented fantasies?

The importance of other kinds of external influences on science, real though they sometimes are, has also been much exaggerated. Here is how Evelyn Fox Keller summarizes the work of physicists: "Out of their interactions with each other, with the public at large, with their own heritage, and with a judiciously culled set of facets of the inanimate world, they have succeeded in producing tools that appear to dissolve nature's resistance to our own needs,"[13] as though Nature played only an incidental role in their findings.

A more extreme instance is that of Paul Forman, whose 1971 article[14] is frequently cited as demonstrating that the development of quantum mechanics, with its relaxation of the classical use of causality, was shaped by the anti-deterministic, anti-rational political atmosphere of post–World War I Germany, in part under the sway of Oswald Spengler's enormously popular book, *Der Untergang des Abendlandes (The Decline of the West).* The author examines in detail the public utterances of prominent German physicists and mathematicians—it was common practice at the time in Germany for major figures at universities to give public lectures on many festive occasions. Not only was "an acausal quantum mechanics . . . particularly welcome to the German physicists because of the irresistible opportunity it offered of improving their public image,"[15] he argues, but more than that:

> [S]uddenly deprived by a change in public values of the approbation and prestige which they had enjoyed before and during World War I, the German physicists were impelled to alter their ideology *and even their science* in order to recover their favorable public image. In particular, many resolved that one way or another, they must rid themselves of the albatross of causality.[16]

Forman concludes: "it seems difficult to deny that the shifts in scientific ideology and the anticipated shifts in scientific doctrine exposed in this paper were *in effect* adaptations to the Weimar intellectual environment."[17] In a similar vein, Lewis Feuer asserts that "without the Munich of 1919," that is, the ultimately unsuccessful revolution that led to a Bavarian Soviet Republic and the resulting counter-terror, "Werner Heisenberg would not have conceived the principle of indeterminacy."[18] The argument is quite specifically that "the program of dispensing with causality in physics was, on the one hand, advanced quite suddenly *after* 1918 and, on the other hand, . . . achieved a very substantial following among German physicists *before* it was 'justified' by the advent of a fundamentally acausal quantum mechanics."[19]

The problem with reading the pronouncements by physicists as pure adaptations to an intellectual milieu without any intrinsic scientific justification is demonstrated by the fact that, although quantum mechanics was not introduced by Heisenberg and Schrödinger until 1925, the specter of acausality had been haunting physics long before that time, at least since Niels Bohr's introduction of his model of the atom in 1913. His theory had no causal explanation for the emission of light by an atom when an electron in an excited orbit descended to a lower one. The same held true for radioactivity, the spontaneous nature of which had been described by Ernest Rutherford as early as 1900. In three papers published in 1916 and 1917, Einstein had introduced the probability concept into the quantum theory, connecting Rutherford's earlier law of radioactive decay with the spontaneous emission of photons, subject to pure chance. The absence of deterministic laws in these cases was already roiling the physics community, and no political or social causes were needed to introduce the idea of acausality into physics. The speeches made by prominent German physicists in the early twenties are more plausibly explained as attempts to adapt their public presentation of the news in physics to the prevailing social and political winds—they were trying to be politically correct. The few scientists who resisted this temptation—most visibly among them Einstein, who had brought the concept of chance into physics in the first place, though it made him extremely uncomfortable—were politically courageous but also, as it turned out, scientifically behind the times.

Heisenberg himself is a witness against Forman and Feuer. In 1919, at the age of seventeen, he served as a volunteer in a para-military unit in Munich, fighting, amidst the general chaos and utter confusion, against the revolutionaries but in odd free moments reading Plato. "I kept wondering why a great philosopher like Plato," he wrote later, "should have thought he could recognize order in natural phenomena when we ourselves could not." Did the loss of the war "mean that all the old structures had to be discarded? Was it not far better to build a new and more solid order on the old?"[20] Plato's well-ordered universe and his own revulsion at the chaos surrounding him remained a lifelong influence on Heisenberg. And yet he became the founder of the very quantum mechanics that enthroned indeterminism, allegedly under the sway of the contemporary political irrationalism and chaos. The argument for decisive social motivations in the development of physics in this case simply does not hold up under scrutiny. It might be argued more plausibly that the conditions in their country made it easier for German physicists to accept the new acausal theory, which some ten years earlier they might have strongly resisted, or that philosophically they were more inclined to do so than French physicists, who had been brought up in a more rational tradition and, in fact, did hesitate much longer before taking to the new physics. It is likely, in any event, that had the Austrian Schrödinger and the German Heisenberg not conceived of quantum mechanics, Paul Dirac, an Englishman, would soon have done so.[21]

Relativistic Social Constructivism

If the sociologists' claims for the existence of extrinsic influences on the thoughts of scientists have some limited validity, the arguments put forward more recently are of a different variety altogether. The last twenty years have seen the rise of the "strong program" in the sociology of science. Here the contention is not only that social or political conditions lead scientists to their ideas, to the specific questions they ask, or to the metaphors they employ, but that all reports of science, both the experimental results and the theories, are conventions that are entirely determined by the scientists themselves: they are all social constructs. In Shapin's words, "it is ourselves and not reality that is responsible for what we know. Knowledge, as much as the state, is the product of human actions."[22] And since we are largely products of society in one sense or another—whether under the influence of society at large or society in the narrower sense of the scientific community—"scientific theories, methods and acceptable results are social conventions."[23]

Some of these critics view science as a "narrative" that has no more claim to cognitive respect than any other narrative, whether it be astrology or folklore. To answer questions about Nature by experimentation is not intrinsically superior to consulting the entrails of sheep. I hasten to add that not all "relativistic social constructivists" go that far, at least directly, in their claims. Moreover, some of them indignantly object to being called anti-scientific. Nevertheless, the writings of social constructivists permeate the discipline of "Science and Technology Studies," and because they originate in universities, they are potentially much more destructive than those of journalists like Appleyard. They are falsehoods that are all the more dangerous for having a small kernel of truth in them—social conditions and psychological states undoubtedly do have *some* impact on the motivations and thought processes of scientists—and therefore appear plausible to the uninformed. Not only are these views without merit, but in the way they are advanced and tolerated they tend to discourage bright young students at universities from being attracted to science. Since, for a great variety of reasons, our society suffers from a shortage of such students, it is important that these arguments be answered and criticized. Even one of the most prominent of the social constructivists, Harry Collins, admits that a "loss of confidence in the scientific enterprise is a disaster that we cannot afford. For all its fallibility, science is the best institution for generating knowledge about the natural world that we have."[24] It's a good idea, then, to look a little more closely at some of the works of these writers, beginning with Collins himself.

After studying first the history of the controversy surrounding the purported detection of gravity waves by Joseph Weber and then the scientific status of parapsychology, Collins concludes that to "a Martian, the world of parapsychology would look like a miniature version of its respectable brother. But parapsychology will never be thought of as proper science on Earth unless it comes to share the institutions of the cognitive life of science."[25] Although Collins seems to imply that the principal reason parapsychology is not regarded as a science is that it lacks institutional support—an idea that I would strongly disagree with—he does raise some pertinent questions concerning the role of experimental expertise in science.

One of the main problems Collins addresses is the indubitable fact that to perform good scientific experiments often requires an enormous amount of skill on the part of the experimenter and an exquisitely constructed piece of apparatus. How do we know that the equipment is working well and that the scientist using it is skillful? Collins insists that "proper working of the apparatus, parts of the apparatus *and the experimenter* are defined by the ability to take part in producing the proper experimental outcome. Other indicators cannot be found."[26] But what is the "proper experimental outcome"? Presumably it is the result of a skillful experiment done with an excellent piece of apparatus. We therefore find ourselves in a vicious circle that Collins calls the "experimenters' regress": the excellence of a laboratory is defined by getting good results, and good results are defined by what is produced in excellent laboratories. His prize exhibit is Weber's claim to have detected gravitational radiation of a strength far above what was expected on the basis of current theory. This claim was ultimately refuted by others, who rested their refutation on the need for detectors and experimenters of such sensitivity and skill that very few places in the world could provide them.

This argument suffers from a lack of attention to the scientific context. If experimenters worked only on a single experiment and their expertise and the excellence of their equipment were judged solely by the outcome of that one experiment, the logic of Collins's "experimental regress" would be unassailable. But this is never the situation in science. Experimenters work on a variety of experiments and their mastery can be judged by their results in many different instances. It is true, of course, that every judgment rests on a comparison with the results of other expert workers, whose skill, in turn, is measured by results that agree with one another. Claims of extraordinary talents that produce unexpected results difficult or impossible to emulate by others should always be—and are—received with great caution, especially if this exceptional aptitude cannot be demonstrated in any way other than by these specific results. That is, of course, exactly the situation in parapsychology, and it is the primary reason why it is not taken seriously. I know of no instances of such claims that have survived scrutiny in reputable science.[27]

The apparatus component of Collins's regress is somewhat more to the point. Experimental equipment is often designed for extremely specific tasks, and it is not always easy to be confident of its sensitivity for its special purpose, particularly when the outcome of the experiment is negative. If a null result is obtained, how can we be sure that this is not because the apparatus was not sensitive enough? Experimenters are, of course, well aware of this dilemma. They deal with it by testing as many components of the equipment as possible separately, in contexts in which the expected outcome is well established. In addition, they estimate, on the basis of the best available current knowledge, what the size of a conceivable positive result would be, and that estimate, in fact, determines the required sensitivity of the equipment. Nevertheless, some doubt often lingers; suspicion may die out only after many repetitions of similar experiments, preferably done by other means, or fade away after the theory the experiment was supposed to test has become firmly established in other ways. The negative outcome of the Michelson-Morley experiment, which dealt a death blow to the then-accepted ether theory, is a case in point. This very difficult experiment required enormous sensitivity of the apparatus and, because doubts of the results remained, was repeated by others, sometimes with ambiguous results.[28]

Collins, like Bruno Latour, whom I shall discuss further on, describes the "negotiations" that go on among central groups in science and insists that these negotiations are essential for establishing scientific results. However, he cautions,

> the outcome of these negotiations, that is, certified knowledge, is in every way "proper scientific knowledge." It is replicable knowledge. Once the controversy is concluded, this knowledge is seen to have been generated by a procedure which embodies all the methodological proprieties of science. To look for something better than this is to try to grasp a shadow.[29]

For him, the relativistic-constructivist aspect of science has its limits, beyond which the program of the sociologists cannot be driven: "The approach we favour is to push the relativistic heuristic as far as possible: where it can go no further, 'nature' intrudes."[30] Some relativists are less cautious.

The Strong Program in Constructivism

In his book *Knowledge and Social Imagery,* Bloor introduced the term "strong program in the sociology of science" (or of knowledge), and Bruno Latour, Steve Woolgar,[31] and Andrew Pickering[32] more or less follow that program. None of them acknowledges any intention of being anti-science, though the result of their deconstruction and their attack on its "privileging" is certainly detrimental to any acceptance of science as an activity aimed at an objective investigation of the world. As will become clear from later remarks, Pickering's protest that he had no wish "to deny reality—in the shape of experimental data—a role in the development of scientific knowledge"[33] has to be regarded as somewhat disingenuous. In fact, these books induce a strange sense of unreality in the reader: words and phrases appear to mean one thing but are intended to mean something entirely different.

For example, in a discussion of mathematics, Bloor[34] quotes the great logician Friedrich Frege: "I distinguish what I call objective from what is handleable or spatial or actual. The axis of the earth is objective, so is the center of mass of the solar system, but I should not call them actual in the way the earth itself is so."[35] Bloor then comments that "the theoretical component of knowledge is precisely the social component,"[36] and concludes, "Institutionalised belief satisfies his definition [of objectivity]: this is what objectivity is."[37] It seems to me it is perfectly clear what Frege meant by "objective," and I don't believe many readers would agree that he meant "institutionalised belief."

At the beginning of his book, Bloor boldly announces that "knowledge for the sociologist is whatever men take to be knowledge."[38] He also attacks the notion that only erroneous beliefs need to be explained in some extraneous way and that correct knowledge and true beliefs require no explanation. Indeed, he regards the relegating of the "externalist" historian to the irrational aspects of science as "humiliating."[39] The "strong program" replaces the "teleological model" (in which knowledge acquisition is directed toward the truth) by a "causal model." To say a belief is socially determined does not imply that it is false, he insists, and he rejects the empirical model, in which there is empirical justification for true belief: *all* belief, including the most rational, needs a sociological explanation. As he puts it, "What I [am] taking issue with [is] every approach that makes 'logic, rationality and truth appear to be their own explanation.'"[40] Treating true or rational beliefs on an equal footing with false or irrational ones is embodied in Bloor's "principle of symmetry." "The strong programme enjoins sociologists to disregard [truth] in the sense of treating both true and false beliefs alike for the purpose of explanation."[41] Nevertheless, there "is little doubt about what we mean when we talk of truth. We mean that some belief, judgment or affirmation corresponds to reality and that it captures and portrays how things stand in the world."[42]

Clearly, there appears to be a contradiction here: if "knowledge is whatever men take to be knowledge," no additional criterion of truth can exist; and if an additional criterion does exist, a sociological explanation for knowledge or true belief is not needed (it may well be interesting to sociologists, but for the scientist it is irrelevant). The key, then, seems to be that Bloor uses the word *knowledge* in an unconventional sense: one can have knowledge of something that is not true. This is confirmed by his further thoughts: "Does the acceptance of a theory by a social group make it true? The only answer that can be given is that it does not . . . Does the acceptance of a theory make it the knowledge of a group, or does it make it the basis for their understanding and their adaptation to the world?—the answer can only be positive."[43] What Bloor ignores, however, is that every group will attach the label "true" to that which it regards as knowledge. Therefore, if he accepts an external criterion of truth—it "portrays how things stand in the world"—he automatically introduces a basis for judging the knowledge of one group as superior to that of others, which he is taking pains to disavow.

A pertinent argument against his insistence that *all* the results of science, and not just the faulty ones, require a sociological explanation has been made by Alan Chalmers.[44] Chalmers offers an analogy: suppose a soccer player sees the ball lying in front of the opposite goal; if he kicks it in, no external explanation is required (he just followed the rules), but if he attempts to eat it, some external explanation is called for, possibly involving his mental health. For Bloor, either action needs to be explained sociologically, and he might well argue that it is sociologically of interest to understand why players follow rules, as well as where the rules originated. For the players and the spectators, however, such questions seem immaterial.

Some scholars in the field of "science studies" spend considerable time in scientific laboratories. They observe the behavior and interrelations of scientists (who, one imagines, at times resent being the objects of observation rather than the observers) as they go about their work, much as anthropologists live in the villages of natives of other cultures and study the day-to-day activities and kinship patterns. Few anthropologists would enter an alien milieu without any prior knowledge of that culture, but these "anthropologists of the culture of science" do not hesitate to observe the behavior of scientists at a seminar or a conference to draw conclusions about why one group rather than another came out ahead in a controversy, indifferent to the reasoning or the logic of any of the arguments—it is sufficient, they believe, to see the social interactions of the players, the display of their forcefulness, self-confidence, social skills, or lack thereof. Sensing that he ought to give at least a minimal justification for feeling that his ignorance of experiments did not disqualify him from judging a controversy in experimental science, Shapin uses as a parallel John Keegan's confession that, though he wrote *The Face of Battle,* he was never near a battle himself.[45] But one may well harbor some doubt whether describing battles without having been in one is the same as judging an intellectual controversy without understanding its content.

Latour and Woolgar base their book *Laboratory Life* on the "field research" by Latour during a lengthy stay at the Salk Institute in La Jolla, California, where he observed the day-to-day activity in a biomedical research laboratory. That he is ignorant of the science pursued there is of no concern to him; if anything, it is regarded as an advantage. The authors explain that their "'irreverence' or 'lack of respect' for science is not intended as an attack on scientific activity. It is simply that we maintain an agnostic position . . . our premise [is] that scientific activity is just one social arena in which knowledge is constructed."[46]

Their central conclusion is that scientific "facts" are nothing but social constructions. As Latour puts it on another occasion, "By itself a given sentence is neither a fact nor a fiction; it is made so by others, later on."[47] The authors maintain that "scientific activity is not 'about nature.' it is a fierce fight to *construct* reality."[48] Indeed,

> "reality" cannot be used to explain why a statement becomes a fact, since it is only after it has become a fact that the effect of reality is obtained . . .
>
> We do not wish to say that facts do not exist nor that there is no such thing as reality. In this simple sense our position is not rela-

> tivist. Our point is that "out-there-ness" is the *consequence* of scientific work rather than its *cause.*[49]

Later, Latour makes the point even more explicit: "Since the settlement of a controvery is *the cause* of Nature's representation not the consequence, we *can never use the outcome—Nature—to explain how and why a controversy has been settled.*"[50] Furthermore, scientists go out of their way to disguise their own responsibility for the phenomena they regard as facts: *"Our argument is not just that facts are socially constructed. We also wish to show that the process of construction involves the use of certain devices whereby all traces of production are made extremely difficult to detect."*[51] Such devices are essential to the operation of science:

> If facts are constructed through operations designed to effect the dropping of modalities which qualify a given statement, and, more importantly, if reality is the consequence rather than the cause of this construction, this means that a scientist's activity is directed, not toward "reality," but toward these operations on statements. The sum total of these operations is the agonistic field. The notion of agonistic contrasts significantly with the view that scientists are somehow concerned with "nature" . . . Once it is realized that scientists' actions are oriented toward the agonistic field, there is little to be gained by maintaining the distinction between the "politics" of science and its "truth"; . . . The negotiations as to what counts as a proof or what constitutes a good assay are no more or less disorderly than any argument between lawyers or politicians.[52]

The argument that a "fact is a fact . . . because it works when you apply it outside science" does not convince these authors. "In no instance did we observe the independent verification of a statement produced in the laboratory. Instead we observed the *extension* of some laboratory practices to other arenas of social reality, such as hospitals and industry."[53] The same laboratory fact works in California and in Saudi Arabia simply because they are the same kinds of laboratory.

> Let us consider one particular statement: "somatostatin blocks the release of growth hormones as measured by radioimmunoassay." If we ask whether this statement works outside science, the answer is that the statement holds in every place where the radioimmunoassay has been reliably set up. This does not imply that the statement holds true *everywhere,* even when the radioimmunoassay has not been set up.[54]

This, of course, is not what scientists mean when they say that scientific facts "work." It is not merely that these facts may be observed under the same circumstances in one laboratory as they were observed in another, but that they become part of an extensive fabric of explanatory statements and ideas—often leading to successful predictions—which we use in other contexts. Facts "work" both within science and without, something that cannot be accounted for by these social constructivist authors.

As Latour and Woolgar are aware, their attempt to demonstrate that the facts observed by scientists are but artificial constructions can be turned against their own "findings." They are, after all, in a situation analogous to the Cretan shouting "All Cretans are liars!" Their flimsy defense is that "neither agonistic nor construction have been used in our argument as a way of undermining scientific fact."[55] But even this can be used against them: if it is true that their own "scientific" argument is immune to the disease of artificial construction, then they cannot claim universal validity of their evidence for the disease by showing purported symptoms of it in only a small part of science.[56] Do these authors expect a reader of their book to come away from it with the conviction that social science is superior to the physical and biological sciences with respect to unconstructed "facts"? They try to defuse such a question by cynically remarking that

> in a fundamental sense, our own account is no more than *fiction*. But this does not make it inferior to the activity of laboratory members . . . By building up an account, inventing characters . . . staging concepts, invoking sources, linking to arguments in the field of sociology, and footnoting, we have attempted to decrease sources of disorder and to make some statements more likely than others, thereby creating a pocket of order.[57]

For Harding, there is no question that, indeed, sociology stands at the pinnacle of all the sciences. She repeats the charge that "physics and chemistry, mathematics and logic, bear the fingerprints of their distinctive cultural creators no less than do anthropology and history."[58] Whereas for "the Vienna Circle, the sciences formed . . . a hierarchically arranged ordering that placed physics at its pinnacle, followed by the other physical sciences, then the more quantitative and 'positive' social sciences (economics and behaviorist psychology were their models) leading the 'softer' and qualitatively focused ones (anthropology, sociology, history)," the feminist reconstructive proposals "reverse the order of the continuum"[59] so that sociology stands at the top of the hierarchy. For the constructivists, sociology not only is first among the sciences in value, but it is necessarily the most fundamental one, underlying all the rest. I have much sympathy for reductionism, but this is reductionism grown rampant, reducing all our knowledge of the universe to a chapter in a text on sociology or social anthropology; physics might as well be taught in a department of folklore.

It is a bit disconcerting that, in spite of his earlier claims, Latour, calling himself a "realist," denies he is trying to "undermine the solidity of the accepted parts of science."[60] He even denies that he is a social constructivist, protesting that he has "written five books to show why a social-constructivist vision *cannot* do the job of characterizing science."[61] Rather, he considers himself to be a relativist when it comes to the production of science but a realist as far as the settled part is concerned. One would have to pervert the sense of the word *realism* considerably in order to reconcile this with his earlier statements. The proper way of interpreting Latour, I believe, is as an extreme—if playful—nominalist, for whom even the words *real* and *social* do not have their usual meaning.

Let us now turn to Pickering, another commentator who denies the distinction between the psychological production of a theory and its scientific justification. As a consequence, he sees an all-pervasive influence of social factors in the acceptance of facts and theories, so that the end result is entirely a social construct. Is it remarkable that Nature is so constituted that it can be rationally understood, or is it to be regarded as an admirable accomplishment of science to have produced a comprehensible and coherent account of this enormously complicated world? Not at all: "Given their cultural resources," he arrogantly announces, "only singular incompetence would have prevented [particle physicists from] producing an understandable version of reality at any point in their history."[62] What Einstein regarded as the most incomprehensible thing about Nature—that it is comprehensible—Pickering views as utterly trivial.

The example discussed at length in his *Constructing Quarks* is the course of events surrounding the discovery, in November 1974, of a new particle, called the *J/psi* (found simultaneously in two laboratories, one of which named it *J* and other, *psi*). This discovery occurred at the beginning of the "November revolution," a period that saw the demise of a theoretical approach called the "bootstrap" method and the rise of the general adoption of "gauge field theory." Other events in high-energy physics involved in this process were the discovery of *weak neutral currents* and of *quarks*, now regarded as the basic building blocks of Nature.

For a period of about ten to fifteen years, quantum field theory, which earlier had been the primary and very successful theoretical tool for understanding the phenomena of particle physics, had been unable to explain the latest observations at laboratories attached to the large accelerators at SLAC (Stanford), Fermilab (Chicago), CERN (Geneva), and others, where particles such as protons or electrons were made to collide with each other at great speeds. For this reason many, but certainly not all, theoretical physicists adopted a new method, the *bootstrap* approach, for calculations and explanations that stayed much closer to the experimental data and required less theoretical machinery than field theory. Instead of explaining the existence of the many observed unstable particles, which are seen as "resonances" in collision experiments (noticeable peaks in a plot of the number of scatterings as a function of the energy of the colliding particles), by starting with the assumption of certain given "elementary" particles and fields, the bootstrap method assumed that all particles, stable and unstable, were on the same footing ("nuclear democracy"). It explained their existence by means of a postulated, self-induced mathematical mechanism that allowed them to pull themselves up by their own bootstraps, *à la* Baron Münchhausen. Fields were regarded as unnecessary, and the only requirement—which, it was hoped, would be sufficient for generating unique answers to all problems of particle physics—was mathematical "self-consistency." For reasons too technical to go into here, the unexpected experimental discovery of the *J/psi* and of the existence of "weak neutral currents" rang the death-knell of the bootstrap approach and led to both the resurrection of the theory of fields in the form of gauge field theory and the prediction of *quarks* as the more elementary constituents of the many particles observed in HEP (high-energy physics) laboratories. Pickering claims, however, this outcome was entirely the result of a different way of looking at the data: "acceptance of the Gargamelle[63] group's novel interpretative practice implied at once the existence of the weak neutral current *and* the existence of new fields for experimental and theoretical research."[64]

Pickering's basic contention is that the physics of the bootstrap approach and the physics of gauge fields are *incommensurable paradigms,* in Thomas Kuhn's terminology. In other words, the physicists in the two camps had no way of communicating with one another, and the experiments that bolstered one paradigm were simply ignored by those who believed in the other. "They would recognize," Pickering argues,

> the existence of different natural phenomena and explain their properties in terms of different theoretical entities. One striking consequence of this hypothesis is that the theories proper to different worlds would be immune to the kind of testing envisaged in the 'scientist's account'; they would be, in philosophical language, incommensurable. The reason for this is that each theory would appear tenable in its own phenomenal domain, but false or irrelevant outside it . . .
>
> Thus the 1960s and 1970s constellations of neutrino experiment and weak-interaction theory were incommensurable: the old and the new theories of the weak interaction were each confirmed in its own phenomenal domain and were each disconfirmed outside it. To choose between the theories of the different eras required a . . . choice [that] cannot be explained by the comparisons between predictions and data . . .[65]

"The old physics," that is, the bootstrap approach,

> focused on the most common processes encountered in the HEP laboratory . . . The new physics instead emphasised rare phenomena . . . New-physics phenomena were invisible by default in mainline old-physics experiments. Old-physics phenomena were invisible by construction in new-physics experiments . . . To attempt to choose between old- and new-physics theories on the basis of a common set of phenomena is impossible: the theories were integral parts of different worlds, and they were incommensurable.[66]

As a participant in the discussions among theoretical particle physicists at the time, I can testify that this depiction of events is wildly exaggerated and quite misleading. Our naïve anthropologist allowed himself to be fooled by the enthusiastic dancing of the natives! The origin of Pickering's claim that the old and the new theories were incommensurable is understandable enough: bootstrap explanations were based on what might be regarded as "formal causes" in Plato's terminology (discussed further in Chapter 8), while field-theoretical ones were fundamentally "efficient causes," based on physical mechanisms. Employing nothing but mathematical postulates, the bootstrap approach lacked all the essentials of physical laws and was always regarded by many theorists as unsatisfactory for that very reason. But Pickering's contention that the experiments themselves and their results were incommensurable is simply based on a misrepresentation of the fact that experimental research in particle physics underwent a change of focus because new theories led to new questions that needed observational answers, which frequently required different experimental techniques. Though Pickering is not ignorant about physics, he clearly went into the HEP laboratories wearing blinders of his own making.

In addition to using words like *incommensurable* where they do not fit the facts, he also impugns the motives of experimenters: "Quite simply, particle physicists accepted the existence of the neutral current because they could see how to ply their trade more profitably in a world in which the neutral current was real."[67] David Cline, the leader of a competing group at Fermilab, had been trying to prove the Gargamelle results wrong, but finally had to throw in the towel. "It is stunning to reread Cline's memorandum of 10 December 1973 that began with the simple statement, 'At present I do not see how to make this effect go away,'" Peter Galison relates. "With these words Cline gave up his career-long commitment to the non-existence of neutral currents."[68] Does this sound like someone who thought he could "ply [his] trade more profitably in a world in which the neutral current was real"?

Summarizing the result of his analysis, Pickering avers that

> the quark-gauge theory picture of elementary particles should be seen as a culturally specific product. The theoretical entities of the new physics, and the natural phenomena which pointed to their existence, were the joint products of a historical process—a process which culminated in a communally congenial representation of reality.[69]

Since "the particle physicists of the late 1970s were themselves quite happy to abandon most of the phenomenal world and much of the explanatory framework which they had constructed in the previous decade," Pickering tells his readers that "there is no reason for outsiders to show the present HEP world-view any more respect."[70] After all, "the world of HEP was *socially* produced."[71] Pickering's extraordinarily sweeping conclusion is that "there is no obligation upon anyone framing a view of the world to take account of what twentieth-century science has to say."[72]

The Hubris of the Sociologists

One can only marvel at the readiness of a sociologist to judge and deprecate the momentous achievements of science in the course of the twentieth century because among practicing scientists, conducting their experiments and erecting this many-faceted and imposing structure, there were strong, even heated, disagreements and reassessments. One of the principal errors the relativistic social constructivists make in marshalling their evidence is to assume that, since in the course of establishing facts or theories scientists engage in heated controversies rather than dispassionate debates, the result can no more be determined by some external reality than the outcome of equally passionate arguments among politicians. No wonder they conclude that the replacement of Newtonian by Einsteinian gravity, or the introduction of quantum mechanics, has no more cognitive significance than the electoral defeat of a Republican by a Democratic candidate or the introduction of Prohibition. A major part of the responsibility for the epistemological relativism that now pervades the sociology of science has to be assigned to the influence of the late Thomas Kuhn, much as he more recently liked to dissociate himself from those disciples whose extreme positions he did not anticipate.[73]

If detectives in a murder investigation have prejudices and preconceived ideas, are we to conclude that the murderer is a social construct, that the person accused is guilty *by definition?* Even a cynic about police work cannot deny that there exists the underlying reality of the deed done by *someone* whom the police are attempting to identify, with or without success. The findings of the detectives may be right or wrong, but if these findings determined guilt by definition, there would be no need for a trial or an appeal. When a panel of experts, after much debate, declares a painting to be by the hand of Rembrandt, their certification may seem to be, for a time, a functional equivalent of the painting being the genuine article; but this is not what we *mean* by asserting that Rembrandt himself painted it. Similarly, the outcome of a passionate disagreement between scientists may or may not be valid knowledge of Nature; but in the end, it is Nature itself that makes the decision, not the social prejudices or the career choices of the participants. It is hard to deny that the temporary belief of many scientists in Blondlot's N-rays (see Chapter 5 for further discussion) had much to do with nationalistic sentiments among French scientists, and that some of the defenders of the pseudo-discovery of cold fusion (also to be taken up in Chapter 5) were motivated by feelings of resentment among chemists over the arrogance of physicists. But does it follow that N-ray machines would now be ubiquitous medical devices if it had not been for the successful resistance by other scientists against the purported evidence for them, or that cold fusion would be a functioning technology if only its "discoverers" had not been sidelined by those with more professional clout? Would caloric and phlogiston correctly account for heat and combustion if only their defenders had had better political or social connections?

What the social contructivists have to say, in the "weak" formulation of their doctrine, may well have some temporary and partial validity for areas of science in flux and turmoil, but little of it applies to established domains, and nothing in the "strong" form is persuasive. To be sure, it is not always easy to tell when a field is settled: the science of geology seemed to be set and fixed for many decades while the idea of plate tectonics was denigrated or ignored. And we should be mindful that the twists and turns of a developing scientific paradigm shift are sometimes buffeted by strong political winds. Nevertheless, science does eventually come to a state in which disagreements are confined to the border areas. If it were true, as these sociologists claim, that the results of the parts of science still subject to controversies are nothing but communal constructs, and the end result of these controversies is determined not by the external world but by social forces, the settled parts too would have no relation to Nature. How could they acquire such a relation if they had none while being built? At what point would social explanations cease to be the sole determinants of the outcome? The "strong program" of social constructivism cannot, without loss of consistency, defend itself by retreating to the parts of science that are in flux.

Certainly, it would be simple-minded to deny the existence of influences between science and the surrounding culture, flowing in both directions. But claiming that profound changes in theories are determined solely by social pressures and political constraints on the scientists adds only obfuscation to the important question of the meaning and validation of scientific knowledge. When all is said and done, by relegating solutions of deep intellectual and fundamental practical problems to the shifting surface currents of social and political contingencies, these commentators arrive at a perverse and grotesque interpretation of science, denying to rational cogitation and logical reasoning, based on empirical evidence, their proper, crucial role. This is the principal reason why most scientists indignantly reject these critiques.[74]

Let me, then, leave the sociologists and turn to a physicist's substantive consideration of the physical sciences. In succeeding chapters I will discuss in turn the purpose of these sciences, the factual evidence on which they are based, and the theories that form their structure. Then I will close with examinations of the quite counterintuitive views of reality they engender, and finally, the very notion of their truth.

3

THE AIM OF SCIENCE IS UNDERSTANDING

"**SCIENCE** seeks to exert power over Nature!" some proclaim, either in anger or admiration; others insist that the purpose of science is, or ought to be, to improve the human condition. For the vast majority of basic scientists, however, the ultimate purpose of their calling is to *understand* the world around them and to explain its workings. "Power," Jacob Bronowski wisely noted, "is the byproduct of understanding."

> So the Greeks said that Orpheus played the lyre with such sympathy that wild beasts were tamed by the hand on the strings. They did not suggest that he got his gift by setting out to be a lion tamer.[1]

It is that powerful offspring of science, technology, which can ameliorate human misery by tempering and harnessing the destructive might of Nature, wielded in the form of pestilences, famines, and other disasters.

To be sure, there are large parts of science, focusing on the systematic gathering and classifying of facts, in which explanation does not appear to be the driving force. Observational astronomy and biology as it was practiced until about fifty years ago are good examples. For millennia most of the efforts of astronomers were concentrated on the collection of more and more detailed information about the heavens, obtained since the time of Galileo primarily by means of the optical telescope and more recently by antennas and particle counters. Some of that information, however, has led to revolutionary ideas and insights—Galileo's observation of the moons of Jupiter with his new telescope and Edwin Hubble's detection of the recessional velocity of distant galaxies had large explanatory consequences. Furthermore, during this century, astrophysics has played an increasingly important part in astronomy, and its success has granted interpretation a much greater role. Whereas the science of biology was, for centuries, dominated by taxonomy—the collection and classification of observational facts—for the last fifty years it also has begun to become more explanatory, especially in the fields of molecular biology and biochemistry. The earlier, fact-gathering ways of these sciences, like the early stages of chemistry, were necessary steps in their development, but a mature science goes beyond the acquisition, description, and tabulation of facts and makes *understanding* its primary aim.

Insisting that understanding and elucidation are the principal goals of science, however, is not, by itself, saying very much. There are other human activities with similar purposes. Moreover, the meaning of these words is not *a priori* obvious. In many cultures and periods of history, myths and tales have served to unravel the mysteries of Nature. In our culture, as well, some explications—most prominently in religion—are far removed from science. Indeed, it is precisely this partial commonality of the aims of science and religion that often leads to clashes, disputes that can be resolved only by clarifying what scientists mean by understanding. The number of religious scientists—Isaac Newton and the astronomer and physicist Arthur Eddington come readily to mind—shows us that conflict is not inevitable after all, between science and religion—the statements of respected scientists like Steven Weinberg[2] and of journalists like Bryan Appleyard[3] to the contrary notwithstanding.

We might choose to explain why the planets move in their orbits by proclaiming that God made the world that way, or we could say that their motion fulfills a mysterious purpose built into the universe, or that it is these special motions which ensure the existence of life on one of them. While any or all of these answers might conceivably be true in some sense and serve as satisfactory explanations in their proper contexts, none of them can be regarded as scientific. They lack the characteristic that is crucial for science as it has developed over the last four hundred years, namely:

Scientific explanations must be part of an intellectual structure that is ultimately justified by objective, public evidence obtained by observation and experimentation on Nature, rather than by divine revelation, scripture, individual personal experience, or authority.

When we want an explanation we ask "Why?"—why does a baseball move the way it does when thrown as a curve ball? Why do earthquakes take place where and when they do? Why does heated sodium emit yellow light? Sometimes, of course, such questions begin with "How?"; there are historians of science, like I. Bernard Cohen,[4] who regard "how" questions as more characteristic of science than "why" questions because "why" seems to ask for a *purpose* and "how" for a *mechanism* or a simple description.

Modern science does not provide explanations in terms of purposes; teleological accounts of natural phenomena that reduce them to final causes are not part of its methodology, because objective answers to such questions cannot be obtained or tested by observation and experimentation. Nevertheless, there can be no doubt many questions that arise in science appropriately start with "why" rather than "how," and it is the answers given by science to these "why" questions that some people, especially philosophers with an Aristotelian bent, find unsatisfactory. If we explain the motion of the planets by means of Newton's laws of motion together with his universal law of gravitation, further queries—*Why* does the gravitational force fall off as the square of the distance? *Why* are the laws of motion just the way they are?—are left unanswered. These kinds of questions are what Einstein had in mind when he declared it his aim to find out "whether God had any choice in the creation of the world," questions some contemporary physicists expect to answer by a "theory of everything." Whether or not these ambitions can ever be fulfilled, scientists have to content themselves in the meantime with more modest goals.

What Needs Explaining?

In his influential book *The Structure of Scientific Revolutions*,[5] Kuhn introduced the notion of the "paradigm," which may be roughly paraphrased as a "research program."[6] The paradigm goes beyond a specific formulation of a theory; it includes a way of thinking about

Nature, about problems and how to attack them most fruitfully. It encompasses such things as what questions arise naturally, what experiments ought to be performed and are likely to yield interesting results, what can be taken for granted, etc. While this notion is very useful for many purposes, it has sometimes been applied far beyond its proper domain, with the implication that scientists using different paradigms in their work have no way of meaningfully communicating with one another (their paradigms are "incommensurable"). In Kuhn's view, scientific revolutions are "paradigm shifts," the acceptance of a new paradigm to replace one discredited. While I agree that a shift of this type occasionally happens, I do not believe the progress of science depends on it to the extent envisioned by Kuhn, and I will use it sparingly.

Nevertheless, when we consider the question, What is it that needs an explanation?, we find that periodic changes in scientists' views of what could be taken for granted and what would require explaining—Kuhn's paradigm shifts—have been among the forces driving the development of science throughout its history. Sometimes phenomena that were regarded as needing an explanation were later no longer thought to demand one, thereby removing obstacles and clearing a path for further progress. Let's take Isaac Newton's introduction of the concept of gravitational attraction as an "action at a distance," for example, an idea that was thought to be utterly repugnant by his contemporaries, and even initially by himself, because it introduced a causal influence without contact. That most physicists now regard action at a distance once again as abhorrent does not alter the fact that by eliminating the need to explain it, Newton made this concept enormously productive for almost two centuries.

More often it was the addition of phenomena previously not regarded as requiring an explanation that provided an impetus for new ideas. In many such cases, the tools needed even to consider asking for understanding did not exist or were insufficient. The heat and light from the sun were, for millennia, taken simply as given or postulated to come from "fire" without further inquiry, until viable theoretical ideas appeared in the nineteenth century. By using the newly developed science of thermodynamics, Hermann von Helmholtz was then able to explain that the heat energy was generated by gravitational contraction—and actually calculate its amount.[7]

During the rapid development of quantum field theory, and especially of quantum electrodynamics, in the 1940s and for years afterwards, the numerical values of the masses of the electron and other elementary particles were not regarded as requiring explanation, nor were such fundamental constants as the electric charges of the electron and proton. In fact, the idea underlying the "renormalization program" of quantum field theory, which made possible precise calculations extremely well verified by experiments, was to use the experimentally measured values of these constants as givens, without further inquiry. Now this attitude has changed, and it has become fashionable among particle physicists to expect theories eventually not to contain any constants that are inserted "by hand" on the basis of experimental results—a "theory of everything" should explain *all* the constants of nature. The freedom of scientists to decide what does and what does not need an explanation helps to make science much more than a collection of propositions simply determined by Nature.

Before approaching the question of what *is* an explanation, we should finally ask *why* we want one at all. There are, of course, many different answers. For some, the purpose of understanding Nature is to exert power, for others it is to relieve human suffering; often these two ends are intertwined. Early astronomers used their knowledge and limited understanding of the course of the heavens for predictions of solar and lunar eclipses primarily with religious intentions, simultaneously greatly enhancing their own prestige. Exerting extensive power over Nature through agricultural methods based on chemistry and biology alleviated much human hunger. From the point of view of basic scientists, however, there can be little question that understanding Nature is its own end and brings its own satisfaction. No external pressure is stronger than the internal force of curiosity.

What is an Explanation?

What, then, is an explanation? It is clear that when we try to explain a state of affairs or a course of events, the words, the metaphors, and the ideas we employ have to be tailored to the vocabulary and the prior fund of knowledge of the listener. We cannot sucessfully explain a phenomenon to a five-year old employing the same words and concepts we use for adults. We cannot speak to a scientifically uneducated person in the terms we use with a trained scientist, or even with a scientist in another area of specialty as with a colleague in the same discipline. The three-year-old whose response to every attempted explanation is another "Why?" shows that what seemed an adequate answer to the adult does not serve for the child. What is more, the meaning of "I understand" may even vary for the same person, depending on the focus of his thought. A mathematician could understand the proof of a theorem in one sense and agree to its correctness and yet claim not to understand, in another sense, "why it works." If a moderately complex proof of a theorem were disassembled into its most elementary logical steps—a procedure mathematical logicians sometimes follow—everyone familiar with the notation would be able to check its correctness laboriously, and yet no one, except possibly a specially trained mathematical logician, would be able to *understand* it. In order to come to grips with the meaning of *explanation* and *understanding,* we have to take into account the prior knowledge and the expectation of the recipient, as well as the complexity of what is being offered.

When we say that we *understand* an explanation, clearly we intend to say that we are able to incorporate it comfortably into the rest of our knowledge, not only without feeling a contradiction but in such a way that we could reproduce the explanation logically from that knowledge. If necessary, part of the explanation might include some new piece of information that we have to accept, but if *all* we were asked to do was to accept the given statement as a new piece of knowledge, we would not feel we were given a satisfactory explanation. When I tell a child for the first time that the light went out because I turned off the switch, my words explain nothing, but they do add a new piece of knowledge. If, after several later experiences of the same kind, the child asks again why, on this particular occasion, the light went out and is given the same answer, she will be content. But as an adult, the same person may not be satisfied with it and may reasonably ask why moving the switch on the wall extinguishes the light.

If understanding means incorporating something comfortably into the rest of our knowledge or beliefs, we must recognize that many disciplines other than science and areas on its borderline as well—

most mythologies and folklores, astrology and witchcraft, for example—mean to serve a similar purpose. The issue of *prediction* now enters as a powerful additional requirement for a *scientific* explanation, which is why physical scientists put so much emphasis upon it. When we think we understand a process scientifically, we ought to be able to use it in order to make a precise and unambiguous prediction; no other mode of understanding has this feature. The precision of a prediction will depend on the manner and degree of comprehension embodied in the explanation, but without prediction a scientist's understanding is regarded as deficient. To understand a concatination of events historically or philosophically entails no such requirement.

An explanation is satisfactory if we are able to reconstruct it logically from our previous knowledge and apply that understanding to circumstances different from those in which it was originally offered. That is why science teachers, to the chagrin of many students in the humanities, put a heavy emphasis on problem solving. In order to demonstrate that she has understood a scientific principle, a student is expected to be able to apply this understanding to situations other than the ones in which she first learned it. A mathematics student does not understand a theorem unless she is able to use it in an unfamiliar context. Neither memorizing nor reproducing what one has seen working in some special instances means understanding.

To turn to a higher level of the same problem, the subject of thermodynamics may serve as an illustration. For some decades in the early nineteenth century, the first and second law of thermodynamics were regarded as perfectly adequate explanations for many observed phenomena involving the behavior of gases and the conduction of heat; in a certain sense, they still are. Yet it became clear eventually to those physicists who accepted the existence of atoms and molecules, still controversial at the time, that the laws of thermodynamics needed a further explanation. If gases consisted of molecules whose motions are governed by Newton's laws, then the laws of conservation of energy and of increasing entropy and all the other laws of thermodynamics ought to follow logically from the well-known principles governing the constituent molecules. Such an explication, to which I shall return further on, was furnished in the second half of the nineteenth century by James Clerk Maxwell, Josiah Willard Gibbs, and Ludwig Boltzmann through the newly formed discipline of statistical mechanics.

In his recent book *The Particle Garden,* Gordon Kane distinguishes three levels of understanding, which he calls *descriptive understanding, input and mechanism understanding,* and *why understanding.* He offers a video cassette recorder as an example: *descriptive understanding* means you can operate it, but you could not repair it if broken; *input and mechanism understanding* means you could repair it without outside help or parts; *why understanding* means you could invent it and design it and make it from raw materials without outside help. This three-level classification serves to distinguish among different kinds of understanding furnished by theories of elementary particles. He concludes that the standard model of elementary particles, the accepted theory at the present time, has achieved the first level, and it would approach the second one if the Higgs boson (the search for which the giant accelerator in Texas had been, in part, designed) were found, because that would allow us to calculate the masses of all the particles in terms of one mass. Whether the third level, which would culminate in a "theory of everything," is really achievable Kane leaves open. For more general scientific questions, this scheme is, I believe, insufficient, and any attempt at a rigid classification of understanding is bound to fail. The human mind works in labyrinthine ways.

Theories as Explanations

Theories are scientists' main tools to explain the workings of Nature, but the word *theory* here does not carry the connotation of conjecture, a connotation that leads many nonscientists to dismiss some part of science by saying "that is only a theory." When Newton emphatically declared *hypothese non fingo,* he was asserting that he did not indulge in speculations. To be sure, some tentative explanations offered by scientists are no more than educated guesses, but these are usually not dignified by the name *theory* until they have acquired at least a reasonable amount of support from factual evidence. The nature and use of such evidence will be discussed later in this book; here I want to discuss only the kinds of theories science offers and their value as explanations.

The fact that theories are intended as explanatory tools, and that the success of that intention depends at least in part on the previous knowledge of the listener, accounts for a phenomenon occasionally encountered in science but often puzzling to nonscientists: the premature theory or discovery. When Daniel Bernoulli proposed in 1738 that the pressure of a gas was caused by the momenta of its fast-moving molecules bouncing against the walls of the container, the world of physics was not ready for the idea. Only after Einstein, in 1905, explained the erratic zigzagging of tiny specks of dust and pollens that had been observed under a microscope, called Brownian motion, as the effect of their random collisions with those very molecules was the kinetic theory of gasses finally fully accepted. The formulation of the laws of heredity by the Austrian monk Gregor Mendel was totally ignored for thirty-five years, until they were rediscovered independently. That "parity" was not conserved—a violation of mirror symmetry—in the radioactive decay of atomic nuclei was revealed in 1929 but looked upon as an experimental error and ignored for some thirty years, because everybody *knew* that Nature was symmetric under reflection. In all of these cases and many more, the fundamental reason discoveries were not recognized as such nor theories accepted was that the community of scientists could not fit them into their conceptual framework and therefore failed to understand them. Understanding could come only when other knowledge had accumulated to such an extent that these discoveries and theories could be accommodated without grating dissonance.

Different Kinds of Theories

To begin the discussion, it is useful to differentiate between a great variety of kinds of theories and the roles they play. Einstein made an important distinction between "constructive theories"—theories that offer detailed mechanisms, such as the kinetic theory of gases—and "theories of principle"—which present abstract principles of wide range, such as thermodynamics and relativity. Only the first, he said, led to real understanding: "When we say that we understand a group of natural phenomena, we mean that we have found a constructive theory which embraces them."[8] The theories of principle he singled out not for their explanatory power but for their wide sweep and their "heuristic value," by which he meant that they had particular "significance for the further development of physics."[9] This is why he titled his revolutionary 1905 paper, in which he introduced *photons* as an explanation of the photoelectric effect, "On a heuristic viewpoint concerning the production and transformation of light."[10]

If Einstein's may be regarded as a "vertical" distinction between theories, there is also what might be called a "horizontal" one—the difference between theories of great generality and others that are meant to be of a more *local* character, in the sense that they encompass only a restricted number of phenomena. Physics is the science that offers the most general theories with the widest range of application to different situations. Of necessity, such general theories will be the most abstract, since they have to be as far removed as possible from specific circumstances. For this reason—its abstractness—physics is regarded as the most advanced of the sciences, as well as the most difficult for nonscientists to comprehend. At this stage of their history, the theories of biology and psychology do not have the sweep of Newton's laws of motion and are always of a much more local nature.

Local Theories

Most theories in science are of the circumscribed kind, that is, they have a restricted range of applicability and are either unconnected to other theories or specific consequences of more general theories. Those of the biomedical sciences, for example, usually are of the former variety, and little attempt is made to connect them to more general ones, primarily because general theories about the nature or causation of diseases at this point do not exist and may never exist. However, there is also a need for local theories in instances where we do have general theories, as in physics.

Local theories based on general laws or principles are usually the ones that allow the most direct confrontation of these principles with observation or experiment. In order to facilitate such a confrontation, scientists must work out detailed consequences of the overarching theory, which is the bread-and-butter occupation of most working theoretical physicists. (Very few of us are in the business of inventing a new theory of relativity every year!) These local consequences of large general laws are often separately called "theories," because their connection with the parent theory either does not form a tight, logical chain of reasoning, or because this chain is very long. Analogously, Euclidean geometry is full of important individual theorems, all of which are logical consequences of Euclid's five axioms but which nevertheless assert "new" things, many of them quite surprising to someone learning them for the first time.[11] In the same sense, local theories in physics that are consequences of larger and wider ones can often legitimately be called theories themselves. What is more, they play important explanatory roles—in many cases much more direct and crucial than that of the general principles—in our understanding of Nature. Nevertheless, it is the general theory that in turn explains the local one.

Emergent Properties

Large-scale systems subject to well-established general laws and principles often give rise to structures that are subject to intricate new local laws; these are sometimes referred to as *emergent* properties.[12] All of the laws of chemistry and biology may be regarded as being of this nature. In physics, an arrow of time in thermodynamics emerges from statistical mechanics, which is based on underlying microscopic laws that have no preferred time-direction. The familiar everyday experience that most occurrences around us are *irreversible*—a video of an egg falling on the floor or of an ink drop dissolving in a glass of water shown in reverse can be easily recognized as running backwards—is explained by the second law of thermodynamics, which requires a very delicate and special local derivation from the general microscopic laws, either classical or quantum-mechanical. This important example of the emergence of new properties is worth a descriptive detour.

In the middle of the nineteenth century, at a time when thermodynamics was an autonomous discipline of physics, Rudolf Clausius invented the concept of entropy, which allowed him to formulate the second law of thermodynamics, originally given both by Clausius and Lord Kelvin, in the simple form: *the entropy of a closed system can never decrease.* As I have already discussed earlier, when the molecular hypothesis became more and more plausible as a theory of the constitution of matter, the need arose to explain the laws of thermodynamics on the basis of that hypothesis. An explanation of the first law, conservation of energy, came easily, because it was an integral part of Newtonian mechanics. But the second law seemed very mysterious. The motions of molecules, after all, are subject to Newton's laws, according to which all processes are reversible—no videotape of moving molecules, if run in reverse, would look in any way odd or would contradict those laws. So what was the nature of entropy, and how could one account for its inexorable increase? The answer was given by Ludwig Boltzmann, who proved that if entropy is defined as the logarithm of the probability of the state of a system such as a gas, made up of a vast number of molecules, *it could never be expected to decrease* as time went on. The second law of thermodynamics thus lost its categorical character and merely stated that for any closed system the entropy increase was *enormously probable*—that is, its probability is extremely close to 1, but not inevitable. While the gap between the previous certainty of predictions and their newly acquired high probability had no practical consequences whatever, it amounted to a large difference in principle, and it took Max Planck, who specialized in thermodynamics and made important contributions to it, many years to accept this new understanding.

The explanation of how Boltzmann managed to wrest an arrow of time from the arrowless Newtonian laws—emblematic of the emergence of surprising new properties from known general theories—rests on two basic facts: first, that the constituents of matter, the molecules, are microscopic, while ordinary, macroscopic containers are filled with a vast number of them; second, that the questions we ask are governed by our concept of causality. We would typically ask, "if we arrange two rooms at different temperatures and then open a door between them, what will happen?" In the natural course of events, the distribution of the air molecules in a room varies and fluctuates, with small deviations from a uniform equilibrium occurring frequently and large deviations very rarely. A graph of these fluctuations would show no preferred direction of time. When we "arrange two rooms at different temperatures and then open a door between them," however, we have artificially set up the system to start in an extremely nonuniform state, at the peak of what in the normal course of events would be a very large and rare fluctuation; it will then *most likely* be followed naturally by smaller variations from uniformity, closer to equilibrium, and thus the entropy will increase: the smaller deviations are more probable. That same large fluctuation, had it not been arranged artificially but allowed to occur naturally, would have been preceded as well as followed by much smaller fluctuations and states of higher entropy. The behavior of the entropy would have been essentially symmetrical with respect to its unusually small value at the top of the large fluctuation; it would have increased toward the future as well as toward the past.

The only reason entropy is extremely likely to increase with time is that we are never in a position of waiting for the two rooms to develop noticeably different temperatures by themselves—this would be a very rare event for which we would have to wait billions of years. But it is not difficult to arrange the rooms artificially at different temperatures, and the question we then ask is "what will happen *after* we open the door?" Furthermore, if the number of molecules were not enormous, the probability of increasing entropy would not be nearly as close to 1 as it is. Thus the local law, the second law of thermodynamics, which serves many important explanatory purposes that even a solution of Newton's equations of motion for all the molecules, if it were practicable, would not fulfill, is seen to be an emergent consequence of these equations.

Consider, as another example, our understanding of the phenomenon of superconductivity. Certain materials have a specific transition temperature below which they completely lose all electric resistance and become perfect conductors. The theory that explains this effect, developed by John Bardeen, Leon Cooper, and John Schriefer forty-five years after its experimental discovery by Heike Kamerlingh Onnes in 1911, is a consequence of quantum mechanics together with the known properties of electrons and atoms. Yet, though every physicist expected superconductivity to be ultimately explicable by quantum mechanics together with known facts, before the BCS theory no one could claim to *understand* it. This would have been true even if it had been possible to program a supercomputer to calculate a solution of the quantum-mechanical equations for such a system (which it certainly was not) and the result had agreed with experiments. The real understanding of this effect required a local theory.

Similarly, an assertion that all biological phenomena are finally reducible to physical principles does not deny the need for and the importance of biological laws. An enormous quantum-mechanical calculation that would "predict" heredity, if it were possible, would explain nothing. The structure of DNA and the double helix, though it is a local theory ultimately based, via chemistry, on quantum mechanics, still remains the proper tool for understanding biological inheritance. And even the structure of DNA is not local enough for an understanding of hereditary specifics—we need the more local theory of the Mendelian laws for that.

Insufficiency of Computation

The fact that being able to make predictions based on large-scale calculations is not sufficient for our understanding of a phenomenon shows the limitation of the utility of computers in science. There can be no question that large and fast computers now play a very important and extremely useful role in science; they allow us to find the numerical consequences of theories with complicated equations, which in many instances would otherwise be practically impossible to solve.

One such case was the investigation of the nonlinear[13] partial-differential Korteweg-de Vries equation, devised in the late nineteenth century to describe the motion of shallow water waves in a one-dimensional channel. Computer calculations in the 1960s led to the discovery of *solitons,* wave-solutions of the equation that might be described as solitary "bumps" colliding with one another; during the collision the bumps are distorted in shape, but after the collision they emerge unscathed, with no change in shape, speed, or amplitude. This quite unexpected behavior of waves, discovered on the computer, clearly called for an explanation, which was, in fact, found soon afterward, and it led to very surprising mathematical developments that went far beyond an analysis of the KdV equation. An unsuspected connection was discovered between the Schrödinger equation (basic to quantum mechanics and linear), on the one hand, and the nonlinear KdV equation (which has nothing to do with quantum mechanics), on the other. The connection can be regarded as a generalization of the Fourier transformation, a mathematical device that is crucial for determining the solution of many linear equations. After this computer-assisted discovery, solitons have been found to occur in many physical and biological contexts. Yet the mere fact, by itself, that they were seen in the results of computer calculations would not have led us to *understand* this phenomenon. Real understanding required further mathematical analysis, which generated a whole new area of applied mathematics dealing with what is now called the "inverse-scattering transform."

Subdisciplines and Intuition

Most sub-areas of physics are, in fact, nothing but local theories whose laws are consequences of larger ones, with the addition of specific simplifying assumptions and approximations. Fluid dynamics and acoustics are cases in point. Largely laid down in the second half of the nineteenth century by Lord Rayleigh, the laws of acoustics that govern the behavior of sound all follow from Newton's equations of motion, together with the laws of thermodynamics (which, remember, are also consequences of those Newtonian laws) and approximations based on the assumption that the pressure variations making up the sound are minute. Similarly, the equations of fluid dynamics are derived from Newton's laws of motion, with specific approximations appropriate for fluids. No one doubts that if we could solve Newton's equations for all the individual molecules, the behavior of every fluid could be described. And yet we still do not *understand* the phenomenon of turbulence! The laws of optics, too, are local consequences of a larger theory, in this case Maxwell's equations for electromagnetic waves (with a bit of quantum mechanics thrown in for the explanation of some special phenomena). But to explain how a microscope works we do not begin with the Maxwell equations.

There are many other instances in physics where local theories are the result of special approximation schemes applied to a larger theory. In particle physics this is a very common procedure. Such local theories—in physics jargon usually called "phenomenological," which should not be confused with the meaning of that word in philosophy—often incorporate, in addition to more or less well justified approximations or intuitive interpretations of a larger theory, local experimental results. Since local theories are often not rigorously derived from the parent theory, their success in confronting experiments, when it occurs, does not directly count as a success of the bigger theory; at best, they can be regarded as a victory for an intuition based upon the latter. In many cases, however, local theories do serve as an important explanatory tool for the interpretation of experimental results that would otherwise appear meaningless, which is why experimenters often like to keep "house phenomenologists" in their laboratory, preferring them to theorists whose work is more rigorously tied to the big theories but is farther removed from directly verifiable results. It is hard to deny that the existence of a large number of local phenomenological schemes in agreement with experimental results has to be counted, in some sense, in favor of the larger theory on which they are based, even though such agreement cannot really be said to corroborate it.

Intuition, a highly prized attribute in physical, chemical, or mathematical form, as well as in other guises, is closely related to such phenomenological interpretations of experimental results. Sometimes physicists profess to believe that their kind of intuition is a quality that gives the best of them some special, direct insight into the inner workings of Nature. While there may be a grain of truth in this, the fact is that the hunches of today's scientists are, of course, very much influenced by today's theories and level of knowledge. Faraday's instinctive insights, great as they were, did not intuit Nature in quantum-mechanical terms. Rutherford's interpretation that the results of the scattering of alpha particles by atoms demonstrated the mass and positive-charge component of these atoms to be concentrated in a very small region at their center was based on classical physics, which in this particular case, by a fortuitous accident, happened to give the same prediction as quantum mechanics, developed more than ten years later. Thus he discovered the atomic nucleus by correct intuition based on the wrong physics.

In its essence, intuition is a very thorough internalization and digestion of current scientific knowledge. It allows a scientist to recognize quickly and reliably which interpretations of new results are correct and to intuit new ones. Constituting a principal part of *scientific understanding,* intuition enables him rapidly to explain new, puzzling experimental results as well as to conjecture innovative theories that later have to be justified with greater care. In practice, highly developed intuition is often much more important than detailed theoretical knowledge. No successful concert halls have been designed by solving the equations of acoustics. In mathematics as well, intuition plays a powerful role in making important advances; many of the greatest mathematicians have made seminal contributions by intuiting theorems that sometimes took others years to prove.[14] Indeed, any new theorem has to be formulated first on the basis of an intuitive hunch before its correctness is fully demonstrated.

The great importance that scientists assign to intuition is justified by the fact that advances always come from new ideas rather than simply from the assemblage of masses of facts. New ideas, to be productive, cannot point in random directions; they must be guided by a sharp sense of what is likely to turn out to be "right" before there is any actual knowledge of it. This is where—in contrast to the broad view of a generalist, who may make intellectual connections and see wider implications but may be less likely to develop the deep insight prized as intuition—the total immersion of a narrow specialist in her subject is enormously beneficial. Such insights guide the theoretical scientist proposing a fundamental new theory on the basis of little evidence, and the experimenter designing an ingenious instrument to address a significant new question to Nature. It is intuition that propels science beyond the edge of its frontier, like a long forward pass in football; but the ball has to be caught—the intuition confirmed—to be counted as a touchdown.

Hierarchies and Reductionism

That the relation between areas of a single science and between various sciences is analogous to the relation between general and local theories implies the existence of a certain hierarchical relationship between scientific disciplines. Solid-state specialists often resent the way particle physicists consider their own work to be the most "fundamental," which in the minds of some implies that it is the most important or the most profound. Setting aside importance and profundity, which are entirely independent of what we call "basic," there can be no question that, in a quite meaningful sense, particle physics *is* more fundamental than material science. As Kane points out,[15] it is the only part of physics that even aspires to a basic understanding at the second or even third level, in his terminology; if it succeeds, he argues, it will indeed be the most fundamental of all the sciences. Furthermore, the search for an understanding of the basic forces of Nature and of the structure of its ultimate constituents does not depend in any way on our understanding of the behavior of large conglomerations of particles. On the other hand, the behavior of particles *en masse* cannot be understood without knowledge of the properties of the individual constituents. When Steven Weinberg writes that elementary-particle physics is more fundamental, he means "that it is closer to the point of convergence of all our arrows of explanation."[16] There is necessarily a hierarchical relationship here.

But the recognition of such a hierarchy does not imply that the behavior of particles in the large can be completely understood as a sum of the behaviors of individual particles. There are, in Nature, collective effects for which the whole is greater than the sum of its parts and entirely new effects emerge at the larger scale. The important phenomena of irreversibility, phase transitions, superconductivity, and superfluidity are examples that easily come to mind. But there is a mystique about such "emergent properties" among enthusiasts for the study of complex systems that far transcends what is justified by the actual situation. The author James Gleick, who introduced the newly fashionable science of chaos to the general public with great success, declared in a talk that "there are fundamental laws about complex systems, but they are new kinds of laws. They are laws of structure and organization and scale, and they simply vanish when you focus on the individual constituents of a complex system—just as the psychology of a lynch mob vanishes when you interview individual participants."[17] But to say that phenomena like superconductivity, which indeed "vanish when you focus on the individual constituents," are subject to "new *kinds* of laws" shows a misunderstanding on his part. It is true that there are collective effects that emerge from a cooperation of a large number of particles, and that the long-range order that underlies many of them can indeed be likened to the psychology of a lynch mob, but the BCS theory that explains superconductivity did not announce a "new kind of law"—its authors shared the Nobel Prize for deriving their theory from the known laws governing electrons and ions in crystals.

Nor does the hierarchical relationship imply that in order to understand the behavior of bulk matter, it is necessary to understand whether electrons and the nuclei of ions making up solids and liquids are "elementary" or are themselves composed of more basic stuff. The standard model of elementary particles is irrelevant for an explanation of ferromagnetism, but knowledge of the properties of molecules and electrons is not. Moreover, as Weinberg observes, "whether or not the *discoveries* of elementary particle physics are useful to all other scientists, the *principles* of elementary particle physics are fundamental to all nature."[18]

The relationship between the sciences of physics, chemistry, biology, psychology, and sociology is similar to that between particle physics and condensed-matter physics. All our understanding of chemistry is based on the properties of atoms, embodied in the periodic table of the elements, which in turn is founded on the basic laws of quantum mechanics, including Pauli's exclusion principle (which explains why the electrons in atoms arrange themselves in shells). This does not mean that an understanding of physics makes all of chemistry a trivial consequence, nor that chemists need to understand all of physics in order to function. There are emergent chemical properties and a specific chemical intuition, which is invaluable to its practitioners, but the laws of physics ultimately underlie all of chemistry.

Biology, like chemistry, once was an entirely independent science, but over the last half-century it has become increasingly dominated by biochemistry and microbiology, which has positioned it ever more intimately in the hierarchy with chemistry. Similar developments may be anticipated for psychology that may cause it to become more and more based on biology, and for sociology, on biology and psychology. At the present time, of course, this prognosis is still quite controversial and bitterly resisted by many psychologists and sociologists, but there is every reason to expect that it eventually will come to pass.

My arguments above constitute what is often referred to as *reductionism*, which in some circles has a bad reputation. Its opponents urge that phenomena be understood in their own terms and with their own explanatory tools; they should not be reduced to parts of other disciplines and modes of understanding. Social phenomena and life, in this view, are *sui generis*, and any attempt to understand them in terms of more fundamental concepts is not only futile but wrongheaded. Many of this persuasion accuse all scientists of being reductionists who are leading us in the wrong direction, where true knowledge cannot be found.[19]

It is indeed the case that science is, in its essence, reductionist—it cannot be otherwise. The simplest way to understand a complicated phenomenon is to reduce its explanation to that of something simpler and already understood. Ultimately, this method cannot be avoided without constantly introducing new *ad hoc* concepts and explanatory tools. Should biologists have been satisfied with the empirical laws of genetics and not have searched for an explanation of its molecular mechanism in terms of DNA? Certainly there are specifically biological laws, but that does not mean they are new creations *ab ovum*. Reduction illustrates directly the intimate relatedness of all parts of Nature, which many of those who oppose reductionism also wish to stress but which they implicitly repudiate. The important thing to recognize, however, is that reductionism does not imply the absence of emergent properties in Nature. Chemistry is full of concepts and theories that go beyond the mere recitation of those in physics; biology has its own local theories that are much more important for biological understanding than simply a knowledge of chemistry and physics, on which it is ultimately based. To predict that psychology will ultimately be grounded in biology is not to deny that psychologists need their own proper tools for understanding; to expect that sociology will someday be founded on psychology and biology is not the same as claiming it is "nothing but" the sum of those two sciences. If we reject these hierarchical relationships, we deny that Nature is whole and interconnected.

I have argued in this chapter that the main purpose of physical science is to explain the phenomena of Nature and to discuss what this means. We should now take a more detailed look at the intellectual instruments and tools used for this purpose.

EXPLANATORY DEVICES

THE phenomenological constructs I discussed in the last chapter are often regarded as *models* rather than theories, implying a certain amount of skepticism about both the reality and the completeness of what is being proposed: "Don't take my proposal too seriously; it's only a model." This skepticism is also evident in the fact that the term *model,* which was always more widely used in other areas of science, is today employed much more frequently in physics than it has been in the past. Even the most fundamental theory of elementary particles generally accepted at the present time is named the "standard model"; no previous theory of such sweep was referred to in that way. But whenever physicists want to emphasize their lack of commitment to the reality of what is described by a theory, or to express their consciousness of its limitations, they simply call it a model. Some carry this agnosticism so far as to refer to all theories as models—most physicists tend to be philosophically somewhat timid—but I will make a distinction, occasionally blurred, between the less weighty models, on the one hand, and laws and theories assumed to come closer to an exhaustive description of Nature or a part of it, on the other.

Models

For more than two centuries, elaborate attempts were made in physics to construct images that would account for phenomena difficult to understand without the help of concrete representations. Such were the *ether models,* built to explain the transmission of a force like gravity through apparently empty space, and later the transmission of electric and magnetic influences and the propagation of light as an oscillatory phenomenon in vacuum. These models began with Descartes, who was the first to give the ether, which he imagined filling all of interplanetary space as vortices of chains of particles, specific mechanical properties enabling the transmission of forces over large distances. John Bernoulli, the younger, won the 1736 prize of the French Academy for his picture accounting for the propagation of light: based on ideas of the elder John Bernoulli, his father, his model proposed that all of space was filled with a fluid containing tiny whirlpools. A century later, when it became clear that the properties of ordinary elastic solids could not explain all the known characteristics of light, James McCullagh invented a new type of imaginary material to do the trick, and the great mathematicians Carl Friedrich Gauss and his pupil Georg Friedrich Riemann, who attached great importance to dynamical models for physical theories, elaborated and extended a similar construction. But it was Lord Kelvin and James Clerk Maxwell who went farthest in devising detailed mathematical and mechanical representations for the material that made up the electromagnetic field. Futile attempts continued until their demise at the hands of Michelson and Morley; Einstein's theory of relativity constituted their funeral.

Most of the ether constructs were meant to be accepted seriously, in some sense, as descriptive of an underlying reality, but there are significant instances in which models that are not meant to be taken literally as theories nevertheless play important explanatory roles. The attempt to understand phase transitions in condensed matter is a case in point. Many materials undergo abrupt changes of some of their properties at certain specific temperatures, the most familiar ones being the boiling and freezing of water. The behavior of a permanent magnet when heated is a less familiar example. At a specific temperature, called the Curie temperature (after the French physicist Pierre Curie, husband of the more famous Marie), it suddenly loses its magnetism. By the middle of the 1920s, the origin of ferromagnetism was, in principle, understood on the basis of quantum mechanics, as was the fact that, owing to the magnetic properties of each electron, atoms with an odd number of electrons acted like tiny permanent magnets. But no explanation had been found for the existence of a specific transition temperature, below which all the atomic magnets in a crystalline domain (an area so small as to require a microscope to see it but containing huge numbers of atoms) are lined up to form a strong magnet, but above which the "long-range order" abruptly disappears. Physicists realized that this would not be easy to explain, since a discontinuous behavior as a function of the temperature could happen only for an idealized system of infinitely many atoms, and the forces among a large number of atoms are complicated. Rather than trying to give an explanation by means of a realistic theory, the German physicist Ernest Ising constructed a model that, on the one hand, possessed all the features regarded as essential for the cause of the phenomenon but, on the other hand, was simple enough to give a study of its mathematical properties some chance of success. The thermodynamic behavior of this now famous *Ising model* of ferromagetism, which used a highly oversimplified scheme for the elementary atomic magnets and their interactions and was never meant to be taken seriously as a realistic picture of an actual piece of solid material, gave rise to a challenging mathematical problem.

The model consisted of markers, arranged in a regular array, that could take on the values $+1$ or -1 (we may think of them as toy magnets capable of pointing up or down, in accordance with the quantum-mechanical behavior of particles of spin 1/2). These interacted with their nearest neighbors only, in such a way that the force of interaction tended to give neighboring markers equal values (i.e., to make the magnets point in the same direction). A powerful mathematical technique of thermodynamics was then applied to determine whether there was "long-range order" (with the magnets in large domains all lined up), and if so, whether there was a temperature above which that long-range order suddenly disappeared. The very simplest such arrangement, of course, consists of all the magnets sitting at regular intervals along a line, in one dimension. In this case, it quickly turned out, there was no Curie temperature below which long-range order existed and above which it ceased. The next step in complexity was a grid-like arrangement (like toy magnets situated in a checkerboard pattern in a plane). After a number of futile attacks, the mathematical problem of this two-dimensional Ising model was finally solved, and the solution showed that it had, indeed, a transition temperature separating long-range order from short-range order. The most physically relevant case, in three dimensions, is generally regarded as too difficult, even for this toy scheme, to have any likelihood of solution.

The seemingly very simple Ising model of the much more complex real situation in Nature is, at this point, the closest thing we have to an explanation of the existence of a phase transition in ferromagnetism, and it may well be the closest we will ever come to one. On the basis of its mathematical success we may say, in a certain very limited sense, that "we understand" the sudden demagnetization above the Curie temperature, though the most familiar phase transitions, the freezing and boiling of water, are not yet in the least understood.

Model building of the kind employed for ferromagnetism is not unusual in theoretical physics. Many of the nonlinear equations in areas of particle physics, in fluid dynamics, and in the theories of complex dynamical systems are much too difficult to solve, and theorists have to resort to simplified models that no one assumes to be realistic or even approximately descriptive of Nature. Nevertheless, if a mathematical model, containing some of the essential features of the actual theory, can be shown to exhibit properties that mirror observed phenomena, it serves as an important tool for understanding those phenomena at least qualitatively. While certainly the most successful theories in physics are those that are able to make very precise numerical predictions agreeing with experimental results, in other situations even the most sophisticated mathematical tools lead only to a qualitative understanding of observations—a reminder that the power of mathematics in theoretical physics goes far beyond the use of numbers. In quantum field theory and particle physics, theorists will often attempt to solve the relevant equations or study their properties in systems with fewer dimensions than are needed for the description of Nature, simply because the realistic case is too difficult. Since mathematically rigorous investigations of the equations of quantum electrodynamics in three space dimensions are still very difficult, mathematical physicists have instead studied them in one and two dimensions, partly as a training exercise and partly as a tool for confidence building; similarly for large-scale simulations of other quantum field theories on lattices rather than continuous space-time. No one believes such solutions are in any way directly relevant to experimental results, but their qualitatitive properties nevertheless serve as useful tools to explain the real thing.

For other sciences, model building is also very common; it is particularly the theory of evolution that invites such efforts in biology. These models are often controversial, not only because their applicability to the real world is not well established, but because the conclusions they lead to may not be based on reliable mathematical reasoning. Look, for example, at the book *The Origins of Order*[1] by Stuart Kauffman, which describes, at great length, a model for the mechanism of biological evolution based on the notion of adaptive complex systems, in which order emerges out of chaos. The controversy engendered by this scheme was aired not long ago in a public debate, organized by the Linnean Society of London, between Kauffman and his former mentor, John Maynard Smith. The exchange did not come close to resolving the controversy,[2] illustrating that it is risky to construct mathematical models too remote from the science they are supposed to emulate. The fashionable disciplines of complexity, chaos, and fractals—as well as that of catastrophe theory, all the rage not so long ago—are full of purported explanations of phenomena ranging from biological evolution to the stock market, from phase transitions to cosmology. After the excitement and the hyperbole have died down and the dust has settled, however, they turn out so far not to have contributed much to our understanding of anything.

One point about the use of models seems quite clear: there is an enormous amount of leeway in their invention, and there can be many different models serving similar purposes. Individual scientists may feel very comfortable with some models and reject others for a variety of reasons, some based on the state of contemporary science or on themes prevalent in their culture. It was surely no accident that the many representations of the ether were all essentially mechanical rather than, say, biological: the science of mechanics was well developed and understood, while biology was not.

Analogies

The manner in which simplified versions sometimes help us understand the implications of a theory is by *analogy,* a powerful general mode of reasoning that may eventually even be successfully incorporated in computer programs.[3] In the instances of the models I have discussed, the analogy with the actual theory is very direct; in other cases it may be much less so and yet serve its purpose very well. Sometimes analogies are drawn between phenomena in quite disparate parts of physics, usually because, despite their apparent differences, they are subject to similar equations. This technique is particularly useful in teaching and in explaining unfamiliar effects to other physicists. In the early days of quantum mechanics, it was quite common for those conversant with it to use analogies from electromagnetic theory in order to help less informed scientists, as well as students, understand it better. Nowadays, many of us have found that in teaching graduate students it pays to use analogies the other way around and to draw parallels to quantum-mechanical phenomena, with which students may already be familiar, to explain electromagnetic behavior. And it is almost unavoidable to mention the lowered pitch of the whistle of a receding train to describe the red shift in the light from stars and galaxies.

Analogies not only help us understand unfamiliar phenomena, but they also serve as powerful stimuli for the imagination in devising new theories and pursuing the search for discoveries. That is why it is always useful for the scientist to have a wide knowledge of various areas of science from which she can draw helpful analogies for a solution to the problem at hand. The Bohr-Rutherford model of the atom was formed in analogy with the solar system—the nucleus as the sun and the electrons as the planets—and though this picture played only a minor role in its scientific acceptance, it greatly helped to popularize the model, however far removed the scales and the underlying physics of astronomy and the atomic world might be. Thomas Young was led to his discovery of the crucial interference effects that nailed down the wave-nature of light by an analogy between light and sound, which he regarded as completely convincing but which his posthumous editor considered "fanciful and altogether unfounded."[4]

In some cases, however, analogies can be misleading. Many analogies in science have been based on Heisenberg's indeterminacy principle, according to which specific pairs of observables of a physical system, called *canonically conjugate pairs,* cannot be simultaneously measured with unlimited precision—if one of them is determined with great accuracy, the value of the other can be known only very roughly. But often these parallels with what is commonly referred to as the "uncertainty principle" are based on a misunderstanding caused by the "explanation" first given by Heisenberg himself and often repeated, namely that the measurement process itself disturbs the measured system in an uncontrollable manner. This so-called explanation, heuristically valuable though it may be, is of very doubtful validity to begin with (as our later discussion of the Einstein-Podolsky-Rosen effect will make clear), and it has been used carelessly by analogy in other disciplines, such as economics and psychology. In these areas, too, measurements produce disturbances in the measured system, but it is incorrect to apply Heisenberg's principle as though it meant nothing more than that attempting to know alters what is to be known. These examples serve as cautionary tales against using analogies in science without having a full understanding of the phenomenon to which the analogy is made.

Heisenberg's indeterminacy principle, and his efforts to make it comprehensible, points to a common practice: appealing to intuition or using ordinary language to explain new, mathematically based concepts or predictions that scientists with a less abstract bent find difficult to comprehend. While this is often very useful, the proffered explanations, plausible though they may appear, are not always correct and may sometimes be quite misleading. Physics popularizations are full of pseudo-explanations that lead readers to think they understand when in fact they have been benevolently deceived.

Metaphors

A form of analogy of a vaguer and more suggestive kind are the *metaphors,* which abound in ordinary speech. Beyond the conventional usage, which is almost impossible to avoid, metaphors are used in physics primarily at an exploratory stage in the genesis of a theory, for pedagogical purposes, or for endowing abstractions with a richer, intuitive context. The suggestive nature of such figures of speech makes them useful for the gestation of new ideas but not for the precision usually needed even in a qualitative explanation. Some commonly used expressions may have a metaphorical origin but are then used in a non-metaphorical sense—physicists refer to "soft photons" and "soft x-rays" in contrast to "hard" ones, meaning not that they are soft as pillows or hard as rocks but that they have low frequencies and their quanta have little energy or have high frequencies and their quanta are very energetic. The three fundamental kinds of elementary particles, grouped into families, are called "red," "green," and "blue," and the theory dealing with them is therefore called "quantum *chromo*dynamics," even though it has nothing to do with colors. One of the properties of these particles is called "charm," though there is nothing particularly charming about them, at least to most people. Names like these are not really intended as metaphors but simply represent the current fashion in physics, which replaced the older custom, still current in many other areas of science, of employing Greek or Latin roots for names of newly discovered properties or entities.

Metaphorical images often serve the purpose of softening abstract concepts. Take, for example, the technical term "scattering cross section," which is a measure of the probability of the deflection of waves or particles by an obstacle. In the very simplest cases, this probability is proportional to the obstacle's geometrical cross section—the larger the cross section of a target, the more likely it is to be hit. But the term is used even when the intuitive geometrical interpretation is quite inappropriate. Another instance is the widely known "black hole," named by John Wheeler, a master at coining colorful terms—there is even a theorem in the general theory of relativity that asserts "black holes have no hair"! What all of these names of abstract concepts have in common is that even though they may appear to be metaphors, they have very precisely defined meanings and their technical use does not depend on the associations they invoke. They are therefore not strictly metaphors.

The metonymic use of the word *clock* in the context of the special theory of relativity represents another kind of figure of speech in physics. It associates the concept of time with the instrument of its measurement, a seminal idea of Einstein's. Thus physicists speak of the "clock of a moving particle slowing down," even though no such particle has a built-in chronometer. What they mean is the much more abstract concept of time of a hypothetical observer moving alongside the particle, imagined as carrying a clock, which we would see running slow.

In sciences such as biology and psychology, where the explanatory structures are less abstract, the use of metaphors is more common. In these and in most fields, the principal utility of this device is generally to offer attempts at explanations to nonscientists. One danger in using figures of speech too freely is that they almost always carry baggage full of intended or unintended associated meanings with them, associations that may distort what was to be communicated. The more serious danger, however, is that their use may be confused with, and take the place of, a real explanation, in which case they may form an obstacle to progress.

Because colorful metaphorical language is such an effective pedagogical device, a great deal of anthropomorphic imagery inevitably seeps into chemistry and biology—imagery which occasionally generates heated controversy. I have just recently come across a lengthy many-sided exchange on an electronic bulletin board over the customary reference in chemistry classes to one molecule *attacking* another. Some of the correspondents strongly objected to such language as being conducive to violence and urged its elimination from the classroom.

Thought Experiments

There is a long tradition among philosophers—famous for their reluctance to get their hands dirty by handling real pieces of apparatus—to use *thought experiments* as a means of convincing opponents or disciples of the correctness of their point of view. These devices are also often used to great effect in science, even though real experiments play a greater role. After setting up an imaginary experiment and showing by logical reasoning, on the basis of the previous knowledge or *a priori* assumptions of the listener, what the outcome has to be, one can expect to arrive at the answer to a puzzling question, or else to conclude that some of the underlying assumptions must be false. Let us look at a few examples.

The sixteenth-century Dutch physicist Simon Stevin solved the following mechanics problem by an ingenious argument. Suppose you drape a uniform chain over an asymmetrical vertical (frictionless) wedge, as shown in Figure 1. Will it slide to one side or the other, and if so, to which? Stevin solved the problem by imagining the wedge held up in space, with the chain closed by joining its two ends at the bottom, as in Figure 2. If you think the chain on the wedge will slide, you have to conclude that the closed chain will keep on sliding

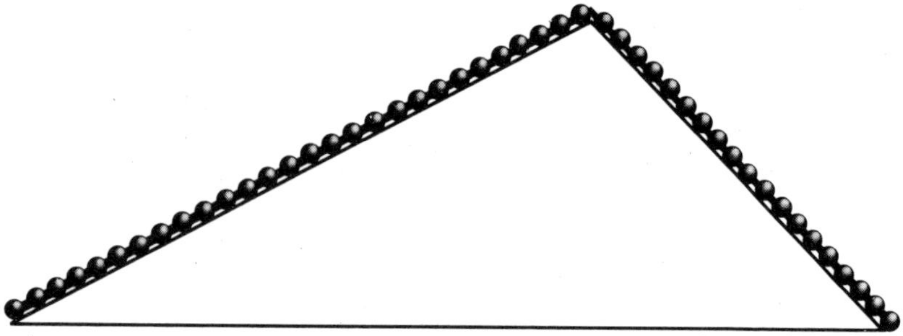

FIGURE 1 The problem set by Stevin: Will the chain slide or remain at rest?

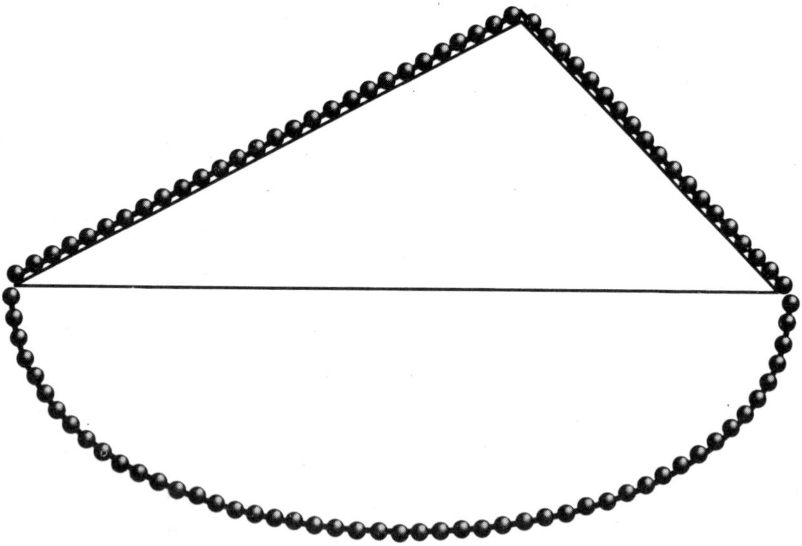

FIGURE 2 Stevin's solution: If the chain moved, adding a lower chain to make a closed loop would result in perpetual motion.

forever; since this is clearly nonsense, the answer has to be that the chain does not move. He regarded this demonstration as so clever that a picture like Figure 2 adorned the title page of his book *Hypomnemata Mathematica*, and his contemporaries showed their admiration by carving it on his tombstone.

Both Galileo and Einstein were masters of the thought experiment and employed the device frequently. Perhaps the most famous of Galileo's was his refutation of Aristotle's assertion that a heavier object falls faster than a light one. He imagined a heavy stone and a light ball, joined together by a rope, thrown from a tower. If the ball falls more slowly than the rock, it must impede the normal fall of the latter and slow it down. But, on the other hand, ball and rock together are heavier than the rock alone and therefore should fall faster than the rock by itself. This contradiction can be avoided only if they both fall at the same speed.

Einstein's recollections furnish us with a third example. Already as a boy he had day-dreamed of running after a beam of light, and it became clear to him gradually that if he imagined himself running along with the beam, at the same speed, he would find himself in the physically impossible situation of seeing an electromagnetic wave standing still, which would contradict the equations, devised by Maxwell, that light is supposed to satisfy. In later years he looked back at these imaginings and considered them to have been the germ of the special theory of relativity.

Einstein's general theory of relativity, the gravitation theory, is also based on one of his thought experiments. Here he pictured himself imprisoned in a free-falling elevator, performing mechanics experiments, throwing balls, etc. Since all the objects in the elevator are falling with the same acceleration, there would be no way of determining that a gravitational field is present and that the laboratory is plummeting down the shaft; for all he could tell, he might as well be moving at uniform velocity in free space, without gravity. This was the birth of the *principle of equivalence,* the indistinguishability of gravitation from acceleration, the cornerstone of the general theory of relativity. Other famous and fruitful thought experiments were designed to elucidate certain paradoxes in the quantum theory, which I will describe in later chapters.

Theories of History

The explanatory devices discussed up to this point are all designed to make us understand phenomena and repeatable trains of events. There are also theories whose purpose is to help us understand the flow of history. All historiography is, of course, in a certain sense an attempt at a theoretical reconstruction. To the extent that it does more than simply recite facts and give the evidence that leads us to believe them—and all good history writing does more than that—a historical text usually relies on arguments of plausibility, psychology, and sociology to explain a course of events. But when that course of events, rather than dealing with the interaction of people, deals with large-scale developments in Nature, the appropriate explanatory instrument that does not rely on oral or written authority is science. All three primary examples of scientific theories of past events—cosmogony, geogony, and biogony[5]—are beset with controversy because they conflict with the accounts offered by various religions, based on authority or individual revelation.

These areas differ from ordinary historiography and may legitimately be regarded as parts of science not only because they use the tools of science for their evidence, but also because implicit in them is the assumption that if the historical accounts and the explanations they offer are correct, the course of events depicted would repeat itself in a more or less parallel fashion elsewhere. While this may also be an unspoken assumption in all other good explanatory historiography, it is much more strongly implied in these cases. Perhaps we are not overly concerned with the way another hypothetical universe would have developed, or even another planet in conditions similar to the Earth's. But when it comes to life itself, our understanding of the course of biological evolution strongly influences our ideas of what living beings elsewhere in the universe might be like and what evolutionary stage they might have reached. Not that we could reasonably conclude extraterrestrial life to be necessarily carbon based like ours, or to have produced copies of *Homo sapiens*—the imagination of science fiction writers certainly roam freely in that respect—but all our estimates of the chances that life has emerged on other planets are based on the premise that our understanding of biogony can be applied to other circumstances. Few historians would venture to draw analogous conclusions from a thorough understanding of the French revolution.

Our increasing grasp of both particle physics and astrophysics has led to a much better understanding of cosmology, and that understanding, in turn, has been used to chart the course of the development of the universe since its beginning, if there was a beginning. For a few years the favorite theory of some cosmologists was that of a "steady state," while others supported the theory of an oscillating universe, neither of which had an initial time. Now, however, the evidence favors the "big bang theory." What we know about the physics of elementary particles, the ultimate constituents of matter, has led to very detailed accounts of how the universe came to be as we now find it, including the formation of the chemical elements and their observed relative abundances in the stars. In these accounts the universe starts from a very dense, hot, and chaotic state in its initial phase, described in Steven Weinberg's well-known book, *The First Three Minutes.* The only part that seems to escape an understanding by means of present theory is the very earliest tiny fraction of a second. As to what came "before" the big bang and what led to it in the first place, physics is silent. Although there are a few physicists who spin elaborate conjectures about "before," there can be no scientific answer to that question if the origin of the universe was a *singularity* in the mathematical sense. (If it was not, we are really dealing with an oscillatory theory or what might be called a "cladistic" steady-state theory rather than a *big bang.*)

It is important to note that the physicist's mode of explanation for the development of a system is always ultimately mathematically based on the use of differential equations, requiring the specification of the system's initial conditions. The differential equation embodies the underlying law and thus the essence of the explanation, while the initial conditions are contingent and left unexplained. Therefore physics does not attempt to explain the state of the universe as it is—it only claims to be able to predict the entire course of the universe, past, present, and future, *provided we know its initial state at the time of the "big bang."*[6]

Since the explanatory tools of cosmogony are those of astrophysics and cosmology, they are subject to all the doubts and uncertainties that trouble our knowledge in those sciences. As we shall see in the next chapter, our knowledge structure in these areas of physics, consisting of a mixture of facts and theories, is still much less firm than in other areas. As a consequence, while there has certainly been considerable progress during the last fifty years in our understanding of the historical development of the universe, we are far from sure that this understanding may not change drastically in the future.

As for our knowledge of the geological history of the Earth, based though it is on much more mundane physics well understood for more than a century, such drastic changes did in fact occur twice, once in the nineteenth century and again not so long ago. The glacial pace of Darwinian evolution required an Earth much older than had been generally believed in the early nineteenth century, and much older than what had been indicated by the Bible: in order for the development of life to have reached its present stage, a timespan of billions of years was needed. But Lord Kelvin, the co-discoverer of the second law of thermodynamics, calculated that if the Earth were as old as billions of years, it should now be much colder than we find it to be, because it would have cooled to a very low temperature by normal heat radiation into space. On the basis of its present temperature, he concluded that the Earth could not be much more than 100 million years old. Using the Earth's heating by the sun, which he assumed to be fuelled by gravitational contraction and doomed to early exhaustion, Hermann von Helmholtz calculated an even younger Earth. These calculations were correct, as far as they went, but two physical facts, discovered at a later date, made them irrelevant: the decay of radioactive elements in the Earth produces enough heat to keep the Earth warm, and the primary source of the sun's energy is now known to be thermonuclear rather than gravitational, therefore much less rapidly consumed.

Alfred Wegener's revolutionary theory of continental drift, which pictured the continents originating in one large land mass that later split into pieces that drifted apart to their present positions, brought about a second drastic change in our understanding of geogony. The new theory was violently opposed by geologists for half a century, primarily because the mechanism and cause Wegener envisioned for the displacement of these large masses of land were ill-founded and confused. Only in the 1960s, when J. T. Wilson and others established the concept of plate tectonics, according to which the Earth's outer layer consists of a large number of thick plates that are constantly in motion, driven by currents in the mantle underneath, was Wegener's idea of the origin of the continents finally accepted. The history of plate tectonics serves as a cautionary tale about the length of time it may sometimes take for an incorrect theory to be driven out and a competing one accepted.

The most famous of the three developmental theories is, of course

Darwin's theory of evolution, which explains the origin of the large number of biological species as the result of the adaptation of animal populations to their environment by means of small random changes, allowing only those most fit to endure the fierce sexual and nutritional competition and to transmit their characteristics to their offspring. The idea, already advanced by others before Darwin, that species changed in the course of history was strongly resisted because it contradicted the Bible. Darwin's notion that evolution, rather than leading to progress and ever greater perfection, as all earlier concepts of species change had proposed, moved toward no discernible goal, was revolutionary. This aimlessness increased the general hostility to his ideas, not only outside science but within as well. Although Darwin amassed an enormous amount of partial evidence for the existence of the changes he envisioned, he did not propose a mechanism that was biologically deeper than the concept of natural selection and the "survival of the fittest." The laws of heredity and their genetic and biochemical explanation, after all, lay in the future. By now, such mechanisms are well understood down to the molecular level, though there is still scientific controversy about details. In addition, evidence for the perpetual changing of species in response to environmental pressures, including new speciation at the present time, is constantly accumulating. When Darwin went so far as to propose that the theory of evolution should apply even to humans, attacks from those who believed in the detailed biblical accounts of the origin of Man increased with fury, and they persist to this day.

The opponents of evolution often contend that it is, after all, "only a theory." While this is correct, and from the point of view of scientists quite innocuous—even the motion of the earth around the sun is, in a sense, "only a theory"—it carries a profound meaning for religious fundamentalists, for whom the account of the Bible has the authority and certainty of scripture, which science can never provide. Here is precisely the point of fissure between science and religion. Since this book deals primarily with the physical sciences rather than biology, I will not pursue the problem of attacks on Darwinian evolution and the enormous amount of evidence in its favor except to point out that, just as cosmogony and geogony rely on all the available tools and knowledge of physics, cosmology, and geology, so biogony must necessarily depend on what is known about biology. The essence of all three of these theories of history is to employ the means of science both for their evidence and for the reconstruction of the course of events.

The Anthropic Principle

The *anthropic principle*[7] is a rather unusual explanatory tool put forth by some respectable physicists, though it is regarded as unscientific by many others, as an attempt to deal with a perplexing question concerning the most fundamental constants in particle physics and cosmology.

Fundamental particle physics, as mentioned earlier, is at this point (and may always remain) unable to explain the numerical values of a large number of basic constants of Nature, such as the electric charge of the electron, the strengths of the strong and weak nuclear forces, the strength of the gravitational force, and the masses of the elementary particles, numbers that determine many properties of the universe and of its constituents. For example, if the strong nuclear force were somewhat weaker, no stable nuclei other than the lightest could be formed, and thus no elements could exist except hydrogen.

In cosmology, it is the gravitational constant that plays a special role, for if the strength of gravity were different, there would be no stars or galaxies. The numerical value of the gravitational constant determines the "escape velocity" of objects in the universe—the speed needed to overcome the total gravitational pull of matter. Now, if the recessional speed of matter created in the explosion of the big bang had been greater than the escape velocity, star clusters such as galaxies would not have been able to develop, unless there were strong inhomogeneities in the distribution of matter to start with. Such inhomogeneities, in turn, would have led to strong large-scale anisotropies—differences in physical properties when measured along different directions—in the present universe, contradicting our observations. (The all-pervading long-wavelength radiation discovered by A. A. Penzias and R. A. Wilson in 1965, whose existence had been predicted by R. Alpher and R. Herman as the cooled remnant of the hot radiation filling the universe at the time of the big bang, has been found to be remarkably constant in all directions, or isotropic.) If, on the other hand, the initial recessional speed had been less than the escape velocity, the universe would have fallen back into a crunch before it had time to become more or less isotropic. Therefore, a universe that was isotropic in the large but that included stars and galaxies could only have developed if there was a delicate balance between the initial speed of matter from the big bang and the escape velocity as determined by the value of the gravitational constant.

Furthermore, life could not exist without stars, or with matter that consisted of nothing but hydrogen; thus the existence of life and intelligence depends on specific values of the strong nuclear force and the gravitational constant. The anthropic principle makes the extraordinary assertion that because this is so, the existence of intelligent life *explains* these particular numerical values of the constants. "Since it would seem that the existence of galaxies is a necessary condition for the development of intelligent life, the answer to the question 'why is the Universe isotropic?' is 'because we are here,'"[8] conclude the physicists C. B. Collins and S. W. Hawking. This strange bit of reasoning clearly needs some explication; there are various interpretations of its logic, each of which has its adherents.

The first formulation of the anthropic principle interprets it as a *purpose:* the existence of humans constitutes the purpose of the universe, and the constants determining the strengths of the fundamental forces were given the values they have so that humans could exist. This reading, which is essentially religious, is not accepted by many physicists, though it has followers among some nonscientists.[9] The second formulation asserts that without intelligent observers, the universe would not exist. It is, in effect, a restatement of the idealism of Bishop Berkeley and has supporters among some physicists who interpret quantum mechanics idealistically (see Chapters 9 and 10). The third formulation, based on a peculiar interpretation of quantum mechanics called the "many-worlds interpretation," states that there are infinitely many different universes, all with different values of the fundamental constants but only one of them hospitable to human intelligence. It is therefore no accident, all three forms of the anthropic principle imply, that we find ourselves in a world in which the constants have the values required for our existence.

A weaker form of the principle declares that the values of the fundamental physical constants are restricted and therefore can be explained by "the requirement that there exist sites where carbon-based life can evolve and by the requirement that the Universe be old enough for it to have already done so."[10] This formulation, though less powerful, is subject to the same basic criticism as the stronger version.

All of these interpretations and uses of the anthropic principle as an explanatory tool have evoked hostility in many scientists,[11] including me. The teleological version of the principle is neither verifiable nor falsifiable and cannot be accepted as a scientific mode of explaining anything; the other versions stand on its head what we mean by an explanation—instead of accounting for the possible development of intelligent life by the structure of the universe, they try to explain this structure by our existence. Would it occur to anyone to explain the extinction of the dinosaurs, which made the further development of mammals possible, by the existence of humans? Such arguments are analogous to the famous ontological proof of the existence of God: that God exists follows from the inclusion of the attribute "existence" in that of "perfection." Explaining the nature of the universe by means of our presence, rather than the other way around, the anthropic principle plays similar games with the slippery concept of existence: since the one thing we cannot doubt is our own being, we take that to be the ultimate source of all explanations. If those adopting the principle find it gives them satisfaction, we cannot deny that it serves a purpose in the way religion does, but it cannot be taken as science; with Freeman Dyson, I am inclined to call it metascience and let it go at that. But the fact that some serious scientists are willing to make use of it shows that even the question of what constitutes a scientific explanation does not have a simple, clear-cut answer, agreed to by all scientists.

There is one version of the anthropic principle that is not subject to the same criticism: it is based on the idea, whose most prominent proponent was Dirac, that all the fundamental constants of Nature are related to the size of the universe and hence vary in the course of time. If that is so, intelligent life could have emerged only during an era when these constants attained the appropriate range of values for this to occur, and we naturally find ourselves in that epoch. In this form, the principle has scientific meaning and is subject to testing, but it does not serve the same explanatory purpose as the other versions. What now needs explaining is the alleged connection between the constants of Nature and the size of the cosmos. There is, however, no evidence for changes in the size of the fundamental constants over long periods of time. The fact (to be discussed further in Chapter 5) that the light spectra of stars and galaxies, emitted hundreds of millions or even billions of years ago, are identical, except for their red shift, to those emitted by the same elements here now is powerful testimony to the unvarying size of the constants, for the detailed spacing of these spectral lines depends critically on the mass and the electric charge of the electron. We must therefore conclude that in its scientifically meaningful form, the anthropic principle is contradicted by the available evidence.

5

THE ROLE OF FACTS

IN the minds of many, the word *science* conjures up images of ever more facts ever more precisely stated, and the collection of these facts is thought to be the rock bottom on which the infallibility and unassailability of science is based. The laws and theories constitute the intellectual framework erected on this factual basis to explain and understand the workings of Nature, or to conquer it. Beyond the theories, in the unfinished upper stories of the building, there are speculations, which the inhabitants indulge in to grasp for new and grander theories, hoping for further understanding.

This picture is, of course, a caricature. Positivists believed they could avoid building the house of science on sand by searching for the simplest statements of fact for the foundation, perhaps admitting nothing but individual sense impressions or readings of measuring instruments: they would design the building in the Bauhaus style, austere and unadorned as an obelisk. This ideal, however, was destroyed by the realization that the ground was full of termites and the atmosphere corrosive. The structure of science is not so simple.

Individual and General Facts

There are facts, and then there are facts. Some of them I will call *individual*—"this rock weighs exactly 10.756 pounds" or "the pointer on the instrument I read on June 25, 1994, at 3:45 P.M. was positioned between the numbers 27.33 and 27.34." Others I will call *general*—"the magnetic moment of the electron lies between 1.00115 and 1.00121 Bohr magnetons," or "at 2.2° Kelvin liquid helium has a phase transition to liquid helium II." Individual facts are generally of no interest at all to science. As Henri Poincaré put it:

> Carlyle has somewhere said something like this: "Nothing but facts are of importance. John Lackland passed by here. Here is something that is admirable. Here is a reality for which I would give all the theories in the world." . . . That is the language of the historian. The physicist would say rather: "John Lackland passed by here; that makes no difference to me, for he never will pass this way again."[1]

For the most part, the only facts of interest to science are those I call general, that is, repeatable and reproducible: these form the touchstones of theories and laws. In order to establish general facts, of course, we have to rely on many individual facts. Measurements must be repeated, and each result constitutes a particular fact, but these are not to be regarded as individual occurrences that are of scientific interest in their own right. That Michelson and Morley performed their famous experiment at the Case School of Applied Science (later called Case Institute of Technology and now, Case-Western Reserve University) in Cleveland in July of 1887, that it took them five days to do it and that they failed to find the ether wind, are singular facts of historical interest, but what is significant for science is that the absence of an ether drift could be, and was, established as a general fact by the repeatability and repetition of the experiment.

There are *some* individual facts that are of interest to science, however, in those areas that have the character of history—cosmogony, geogony, and biological evolution—and in the fields that deal with establishing the constitution of the universe—astronomy, astrophysics, and cosmology. In the first three disciplines, scientists are indeed interested in the analogue of "John Lackland passed by here," because their primary focus is on exactly how the universe, the Earth, and life developed. Cosmologists are in a position similar to that of explorers in search of new territory, but since they can't get there in person, they have to employ indirect methods utilizing general facts and scientific theories for their inferences.

That all the visible stars are made of the same stuff as the Earth, which is at the basis of our understanding of the make-up of the universe, is an individual cosmological fact, but we depend on general facts for our body of knowledge about it. The light emitted by the atoms of each element has a very characteristic color and, when decomposed into its various constituent frequencies and recorded on photographic film, exhibits a large number of *spectral lines* uniquely associated with that element, like a fingerprint. This spectrum allows us to identify the element that emitted it—even when only the minutest amounts of that element are present—and "spectral analysis" has become an important practical tool in chemistry. When the light reaching us from distant stars is spectrally analyzed, we find that all the spectra in it can be associated with the same kinds of chemical elements that we know on the Earth.[2]

The general lack of interest that scientists have in individual facts carries consequences that are rarely discussed. As mentioned earlier, the second law of thermodynamics pronounces certain physical processes to be irreversible. If a door is opened between a warm room and a colder one, the two rooms will necessarily come to equilibrium at a common intermediate temperature, while the entropy of the air in the two rooms has increased. In the subsequent, more fundamental setting of statistical mechanics, this increase of entropy became enormously probable, but it was no longer a certainty. Suppose now that someone observes a strange occurrence: he opens a door between a hot room and a colder one and to his astonishment finds the warm room getting hotter and the cool one colder. This is perfectly compatible with the laws of physics, but the probability of its occurrence is so small that it cannot be expected to have taken place even once since the beginning of the universe. Because it cannot be repeated, it has the status of an individual fact that has no scientific significance and therefore is of no interest to physics. We can conclude, then, that even though there appears to be a large philosophical difference between a law that rules out an occurrence altogether and one that gives it an extremely small probability, not only is there no practical difference between them, there is also sometimes no scientific difference.

The difference between saying something *cannot* happen and saying its probability is extremely small does become significant, however, when there are many repetitions of the occasions on which it might happen. Take, for example, the origin of life. Suppose that the probability for RNA to form by happenstance within one year in the soup of organic molecules on a planet such as the Earth in its early stage of development were estimated to be one in a billion billions and that there were also an estimated one billion such planets in the universe; then there would be a sizable probability that RNA would, indeed, be formed by chance in a billion years on at least one of these planets, and that one we call Earth. You can see from this example that the occurrence of an event of extremely small probability does not always have the status of an individual fact with no scientific interest.

Facts Summarized as Empirical Laws

Scientists often group a number of similar and apparently related facts into an empirical law. For example, a voltage applied to an electrical circuit produces a current; when the voltage is doubled, so is the current. More detailed experimentation led to Ohm's law, which states that the electric current in a conductor is equal to the applied voltage divided by the resistance. Boyle's law, mentioned in Chapter 1 (which was really discovered by Richard Townley by perusing the data Boyle had published),[3] might be cited as another example in physics. Such laws have a character that places them on the borderline between general facts and theories; they are local laws expressing *relations* among facts established purely empirically. Like Ohm's law, they usually are not universally valid but may be of great practical importance in many applications.

In other sciences empirical laws are also extremely common. Physiology has the all-or-nothing law, relating a stimulus to its response in excitable tissues like the heart muscle—below a certain threshold a stimulus produces no response, but above it the response is maximal. In psychology there is Fechner's law concerning the sensory ability to distinguish barely noticeable differences in stimuli. And in economics there is the law of diminishing returns.

Laws of this kind are the simplest attempts by scientists to find relations between groups of general facts, and, if these laws are to be understood, they always need a later explanation by a general theory. We see this at work in the case of Ohm's law, where the explanation was given in terms of the motion of electrons in a piece of metal, long after the law was originally proposed. The general theory that explained Boyle's law was statistical mechanics, governing the large-scale behavior of the many molecules that constitute a gas. It should be noted, however, that most other sciences have not yet reached the stage at which their local empirical laws can be understood on the basis of a general theory.

Establishing Facts

I have mentioned before that there are relatively young areas of science relying primarily on establishing and classifying facts, concerned more with taxonomy than with explanation. Until half a century ago, biology was still at that early stage and had been there for a very long time. The categorization of genera, species, and varieties of living organisms is an activity that falls short of explaining anything—at least in the sense in which we use that term now;[4] it occupies a place intermediate between the gathering of facts and theorizing. Some fifty years ago, a fierce struggle ended the hegemony of the taxonomists, and biology took the next step toward becoming explanatory in the modern sense.

What makes one fact more interesting than another is its relation to some explanatory scheme: either it fits into a theory and corroborates it, or it contradicts a theory and leads to an "anomaly." If it does neither, nor promises to, it remains a "mere fact" and is quickly forgotten. Sometimes a fact may simply be a puzzle; it may "smell" interesting but cannot be placed immediately into a theory or held in argument against one. Finding and establishing an important new fact that can be situated in a theoretical context constitutes a *discovery*. In some instances, like the discovery of radioactivity by Henri Becquerel or the bacteria-killing property of penicillin by Alexander Fleming, this happens by accident. (Note that the very names given to the discoveries of the clouding of a photographic plate and of the unusual behavior of a mold place them in an explanatory scheme.) However, this is rare—the importance of serendipity in science is sometimes exaggerated; more often, discovering a fact—designing the successful apparatus for making a discovery and convincing the scientific community of its validity—takes great ingenuity and a concerted effort. (Sociologists of science are usually more interested in the activity of persuading others than in the discovery itself, while textbooks tend to ignore it.) It might be instructive to take a look at an example of a clever experiment of fundamental importance.

We now know that electrons have a specific electric charge, all of the same magnitude, whose numerical value, 4.80×10^{-10} electrostatic units, is now listed in every textbook of physics. This basic fact was established by Robert Andrews Millikan in his famous oil-drop experiment (Figure 3)—conducted over a number of years and published in 1913—against the fierce opposition of a competing physicist who was convinced his own experiments proved otherwise. The idea of Millikan's experiments was to allow tiny, electrically charged oil droplets to fall in the space between two horizontal, electrically charged metal plates. The electric field between the plates exerted an upward force on the droplets proportional to their charge, and when he adjusted the field so that the electrostatic force exactly balanced its weight, the droplet hung suspended in midair. Knowing the strength of the electric field needed to achieve this equilibrium allowed Millikan to calculate the electric charge on each individual drop. This difficult balancing act led him to conclude that the electric charges on his drops always came in small, integral multiples of a single fundamental unit, the numerical value of which is now known to be one of the most basic constants of Nature.[5] Most students of physics today are asked to repeat this experiment, which requires great diligence and patience, and remember it as a frustrating exercise. The efforts of Millikan's competitor, a man of much wider reputation, are forgotten. Even the most reputable scientists, however, can sometimes delude themselves into thinking they have made a discovery, as the following examples demonstrate.

Pseudo-Facts

In 1903, not long after the discoveries of cathode rays, x-rays, and Becquerel rays (radioactivity)—when rays, in fact, were the rage in science—the prominent French physicist René Blondlot announced the discovery of a new kind of radiation, which, in honor of his native city of Nancy, he called "N-rays."[6] For several years, this discovery, published in the most reputable scientific journals, was discussed in some 300 articles by over 100 scientists. Even though many other physicists had a very hard time replicating Blondlot's experiments (because, Blondlot declared, they lacked the requisite skill), over forty French physicists claimed to be able to detect these rays emitted by many different kinds of substances, including the human nervous system. In Blondlot's apparatus, the N-rays were refracted by an alu-

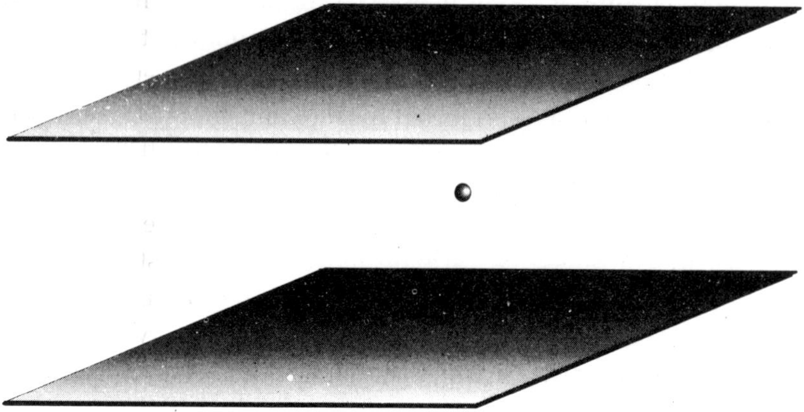

FIGURE 3 Schematic picture of Millikan's electrically charged oil drop suspended between two charged plates.

minum prism and observed by faint spots on a screen placed in front of it. The discovery, however, evaporated when the visiting American physicist, Robert W. Wood, during a demonstration, surreptitiously removed the aluminum prism, which supposedly directed the N-rays toward the screen, while Blondlot, unaware of the prism's removal, kept on seeing the alleged effects of their presence anyway. Clearly, what Blondlot had thought was caused by his new rays was instead produced by various other factors, and he, as well as many of his followers, were the victims of an illusion.

The pseudo-discovery of *polywater* is another example of a well-publicized event gone sour.[7] Some thirty-five years ago, the Russian chemist Boris Derjaguin announced he had discovered a new, anomalous kind of water, which he had condensed from vapor in fine glass capillaries. It had an unusual spectrum and uncommonly high viscosity, it weighed 40 percent more than ordinary water, and instead of freezing it went into a glassy state at $-30°$ Fahrenheit. What he had found seemed to be a hitherto unknown form of H_2O, but it could be made only in very small quantities. His announcement stimulated an enormous research effort all over the world, stretching out over almost ten years and producing hundreds of publications, even a detailed molecular model for the alleged new water polymer. The revolutionary substance was eventually found to be water contaminated by all sorts of impurities, including probably the sweat and elbow grease of the hard-working scientists, and it was these contaminants that accounted for its unusual properties. The episode was a classic case that demonstrates the importance of the scientific requirement that a general fact be checked by others. When the replication did not work, the purported "fact" disappeared.

The "discovery" of gravitational waves, already briefly alluded to in Chapter 2, furnishes a third example. One of the striking differences between Newton's theory of gravity and Einstein's is the prediction of gravitational radiation by the general theory of relativity. Heinrich Hertz's discovery of electromagnetic waves had helped to confirm Maxwell's theory, which had pointed to their existence; now the detection of gravity waves would serve as a brilliant corroboration of Einstein. Just as electromagnetic radiation is emitted by accelerated electric charges, so gravitational radiation should be emitted by moving objects—any massive body struck by such a wave of rapidly varying gravitational force will accordingly be made to vibrate. The catch is that this radiation is predicted to be exceedingly feeble and the ensuing oscillation not only very difficult to detect, but also hard to distinguish from vibrations caused by other random disturbances. There is no chance of measuring the weak flux of gravity waves coming from a terrestrial object whose motion would be controlled by an experimenter; only the rapid motion of a celestial body of gigantic mass, such as a supernova or a black hole, could produce enough radiation to enable its detection by the most sensitive antenna. And since the putative source of the waves would not be under the control of the observer—it could not be turned on and off at will—any positive observation would have to be carefully distinguished from reactions of the detector to other causes.

In 1969, the physicist Joseph Weber at the University of Maryland undertook to attempt this difficult feat by constructing an antenna consisting of a large aluminum-alloy cylinder, weighing more than a ton. The cylinder, suspended from a thin wire and encased in a vacuum chamber, was isolated as much as possible against all extraneous disturbances, its internal vibrations to be measured by supersensitive crystals, amplified, and recorded by a pen on a sheet of graph paper. Since the natural thermal oscillations of the atoms in the antenna—its great weight notwithstanding—would produce enough vibrations to register on this sensitive apparatus, any significant gravity-wave signal would have to be picked out from a constantly present background "noise" shown on the jittery graph. After lengthy observation, Weber announced he had found several large signals per day that clearly stood out against the random background. Over the next several years he added refinements, such as positioning a second similar antenna six hundred miles away, whose output was temporally correlated with the first, to be sure that the arrival of broad gravity wave-fronts hitting the Earth would register simultaneously on both. The signals he saw even appeared to show a twenty-four-hour periodicity, suggesting that the radiation came from a specific extraterrestrial direction. He had finally discovered what Einstein had predicted, or so it seemed.

The problem with Weber's discovery was that the size of his signals indicated a strength of gravity waves that was enormously much greater than could possibly be accounted for by any known source in the sky. In addition, by 1972 several other laboratories had gotten into the act, building their own gravity-wave detectors of much greater sensitivity, and none of them saw these large signals. Furthermore, controversy surrounded Weber's statistical analysis underlying his sifting of signals from the ubiquitous noise and the correlations between peaks on the graphs from his two distant antennas. After years of debate, the dispute ended with the consensus that the strong gravitational radiation Weber had seen was mostly a statistical artifact and had no reality.[8] As for weaker gravitational waves from known potential astrophysical sources predicted by Einstein, they are still below the threshold of observability even by the most sensitive antennas constructed. So far, they have been detected only indirectly, by a gradual increase in the rotation rate of binary stars, caused by the energy loss due to the emission of gravity waves.

More recently, a loudly advertised "discovery" of great potential industrial importance turned into an embarrassing non-event. In 1989, two reputable physical chemists, the American S. Pons and the British M. Fleischmann, announced at a press conference at the University of Utah that they had discovered *cold fusion* of deuterium nuclei in electrolytic cells with palladium cathodes and heavy water (D_2O) electrolyte. Their claim rested on the reported observation of a much larger amount of heat than they could account for by any known chemical reaction; they therefore interpreted the heat as evidence of nuclear fusion, achieved without the need for the very high temperature that is usually assumed to be required in order to give the positively charged deuterium nuclei enough energy to overcome the electrostatic repulsion between them and enable them to fuse. What physicists had tried to accomplish with huge machines at great expense, two chemists apparently had done in small flasks on a tabletop. However, when others tried to duplicate their experiment, at laboratories all over the world, almost no one succeeded, though a few claimed to corroborate the discovery. The purported evidence for the nuclear, rather than chemical, nature of the process, namely the detection of accompanying neutrons and gamma rays, was always vague and could not be verified. Though very skeptical of the claimed effect, physicists were at first generally cautious, but soon they repudiated it; many chemists, on the other hand, hailed the announcement, welcoming a discovery that showed they could do very cheaply what the physicists had tried to do for many years at immense cost. After a while the clamor died down and there is no remaining evidence that what Pons and Fleischmann had seen was anything but a normal chemical process. Nor was the heat produced of any industrial significance. The "discovery" had evaporated.[9] Let us return, then, to facts that are more secure.

Theory-Laden Facts

The facts upon which scientific laws are based are almost never established in pristine isolation; rather, in one way or another, they usually depend on these very laws: they are *intertwined* with them. As philosophers of science have frequently noted, many so-called facts are "theory-laden," which is to say that the way in which they are demonstrated, and even their very meaning, depend on theoretical interpretations. Examples abound.

The determination of the electric charge of the electron described earlier relied on the use of a number of theories. Millikan calculated the electrostatic force between the metal plates by using the known laws of electrostatics. In order to determine the weight of each oil drop, he needed to find its size, which he did by turning off the electric field and measuring the constant terminal velocity of the drop when falling in air. The dependence of that final speed on the diameter of a spherical drop, in turn, follows from the known laws of gravity and air resistance. As in many other cases, the instrumentation for the detection of an experimental fact utilized other theories and interpretations of other facts.

Here are some further instances of basic scientific facts: light moves with the finite speed of 3.00×10^{10} cm/sec; almost all of the mass of the atom is concentrated at its center, in a nucleus whose diameter is about 10^{-4} times that of the atom; many of the fundamental particles discovered during the last fifty years are unstable, and their known masses and half-lives are basic facts of Nature; the distance to the Andromeda Galaxy is about two million light years; the age of the universe is about 15 billion years. How did we come by this information?

The first determination of the speed of light, at the time thought to be infinite[10] but now known to be one of the most fundamental constants of Nature, was performed by the Danish astronomer Ole Christensen Römer in 1679. His inference was based on his observation that the time intervals between successive eclipses of Jupiter's moon Io varied by as much as ten minutes, depending upon the positions of Jupiter and the Earth in their orbits around the sun. These variations, he reasoned, were caused by the fact that, when the Earth was farther away from Jupiter, light took that much longer to travel here. From the known laws governing the orbital motion of Jupiter and the Earth, he was able to estimate the speed of light to within 25 percent of what is now known to be the correct value.

It was Ernest Rutherford who, in 1911, discovered that atoms have a very small nucleus at their center, in which almost all of their mass and their positive electric charge are concentrated. He based this conclusion on experimental observations produced by pointing a beam of alpha particles (doubly ionized helium atoms and therefore positively charged) at a thin, gold foil and finding a fair number of them deflected back. If he assumed the accepted model of J. J. Thomson, which pictured the atoms as more or less uniformly positively charged balls in which the much lighter negatively charged electrons were embedded like raisins in a pudding, his observations strongly disagreed with the prediction based on electrodynamics. Such an atom would hardly ever bounce the alpha particles in the backward direction. On the other hand, Rutherford could account for the observations very well if all the positive electric charges of the atoms and most of their mass were concentrated in a tiny region no larger than about 10^{-12}cm in diameter, or only one-tenthousandth the size of the atom. This inference was therefore based in considerable detail both on Newton's laws of motion and on the laws of electrodynamics.

There is some irony in the fact that these laws are now known to be inapplicable to atomic systems and that a correct calculation of the scattering observed by Rutherford had to be based on quantum mechanics, not yet discovered in 1911. By a remarkable coincidence, however, the detailed angular dependence of the so-called Rutherford scattering happens to be the same when calculated classically or quantum-mechanically. If it had not been for this fortuity, the discovery of the atomic nucleus might have had to wait for some years.

On what basis do we know that the half-life of the tau meson is 4.6×10^{-13} sec? The lifetime of this unstable particle is inferred from a knowledge of its velocity and the length of the track it leaves in a detection device, such as a bubble chamber or a spark chamber, from the point of its creation to its decay into other particles. The ratio of the measured track length to its velocity, however, is very much larger than 4.6×10^{-13} sec. To arrive at this shorter lifetime, the theory of relativity has to be taken into account, because these particles, created in high-energy collisions, move through the detection chamber almost as fast as light. According to relativity, the clock (metaphorically speaking) of a moving system appears to run at a slower rate, and the half-life of the tau meson therefore seems to be stretched out, leaving a longer track in the stationary chamber than it would in a laboratory that moved along with it. So the measurement of one of the fundamental properties of a highly unstable particle depends on the use of the theory of relativity. In other instances, in which the lifetime of a particle is so short that no trail is visible, detection of the half-life is even more indirect.

Now let us look at the method by which the distances to faraway galaxies and stars are determined. Stimulated by Willem de Sitter's solution of Einstein's equations, which showed distant galaxies receding at a rate increasing with distance, as well as by Vesto M. Slipher's research on spiral nebulae, E. P. Hubble made the great discovery in 1929 that the universe was indeed expanding and that there appeared to be a constant ratio between the recessional velocity of a galaxy and its distance from us. In other words, the speed with which galaxies recede from us is proportional to their distance;[11] this ratio or universal constant of proportionality is now called Hubble's constant. Not only does this law have profound implications for the structure and development of the universe, but once its numerical value has been determined from one set of galaxies, it permits us to infer the distance to any other galaxy from a measurement of its recessional velocity.

The speed with which a galaxy recedes from us, on the other hand, can be easily determined from its red shift, the amount by which each spectral line of the light emitted by its atoms and observed on Earth is shifted toward the red end of the spectrum. This red shift is interpreted as a Doppler shift analogous to the sudden drop in the pitch of the whistle of a train as it passes and recedes. The decrease in pitch is proportional to the recessional speed, and hence, if the Doppler shift is the cause of the red shift, the velocity of the fleeing galaxy can be easily inferred from the numerical value of the shift of all the lines in its light spectra.

But Hubble had to find the distance to a galaxy directly, which is more complicated. The first thing to note is that astronomical distances are usually given in light years, that is, the distance traveled by light in one year—9.45×10^{12} kilometers. The reliability of the conversion from light years to kilometers, however, depends on the special theory of relativity, according to which the speed of light is constant and independent of the speed of the observer.

The essence of a direct measurement of an astronomical distance is a comparison of the apparent brightness of a stellar object to its intrinsic luminosity; application of the inverse-square law for the light intensity then allows us to use that comparison to infer the distance. But how can we learn how bright the star itself is? The traditional method, used by Hubble himself, utilizes the special properties of certain pulsating stars known as Cepheid variables. Since, by a well-accepted law, the period of the Cepheids' pulsation bears a strict relation to their intrinsic brilliance, observation of the variation in their periods allows us to figure out that luminosity, and their apparent brightness enables us to infer their distance. If a galaxy contains[12] such a Cepheid variable, we can therefore tell its distance. (On the scale of these intergalactic distances, galaxies may be regarded almost as points.)

The only problem with this procedure is that Cepheids are not very luminous and hence, if they appear in very distant galaxies, they cannot be seen or identified, and hence the distance to such galaxies cannot be determined by this method. There is, however, another distance indicator that can be used for farther distances, namely supernovae of "type II." The development of these supernovae is believed to be sufficiently well understood to permit the use of computer models of their history, enabling astronomers to calculate their true size from a number of observable data. Distance determinations for galaxies by means of these supernovae are not yet regarded as quite as reliable as those by means of Cepheids, but they have the advantage that the supernovae are more brilliant, which allows distance computations for galaxies much farther away, where the identification of Cepheids is ambiguous.

It should be clear from the foregoing account that when we speak of the distance to a certain galaxy as known, the purported fact depends very heavily on many theoretical assumptions. This became particularly apparent in recent controversies.[13] One group of astronomers employed "type Ia" supernovae for the distance determinations, arriving at the usually accepted value of the Hubble constant, and other groups used "type II" supernovae, as well as Cepheids so far away that their identification may have been ambiguous, arriving at a Hubble constant about 80 percent larger. A larger Hubble constant implies that all the galaxies we see are much closer to us than previously thought. What is more, the Hubble constant, together with an estimate of the total amount of mass in the universe, is also used for an estimate of the age of the cosmos. Using the larger Hubble constant, cosmologists concluded that the universe is much younger than had been believed, in fact, younger than many individual stars had been determined to be! The resulting puzzle, exactly analoguous to one that arose more than fifty years ago, remains, so far, unresolved.

As we can see from these examples, the extent to which the facts of science are theory-laden varies greatly. Some facts are relatively pure and independent of theoretical assumptions, others are heavily "contaminated." Our knowledge that the stars are made up of the same chemical elements found on Earth is based only on the identification of their electromagnetic spectra, the same, except for an overall displacement, as those observed here on Earth. Save for the correction made to take care of the red shift, this reasoning does not use any physical laws. (The inference drawn from the red shift that the star is rushing away from us, on the other hand, is, of course, theory-dependent.) At the other extreme are many other facts of astrophysics and cosmology, established very indirectly and utilizing an extensive scaffold of theoretical reasoning. As a result, they change relatively frequently when some of the assumptions are found wanting.

In general, it is fair to say that almost every experimental result used to corroborate or disprove a theory can serve that purpose only after being interpreted by means of either another local theory or another part of the same general theory. In most instances, the daily work of experimental scientists consists of playing off one local theory against another, in the sense that one of them is utilized for the interpretation of the outcome of the experiment, and the resulting "fact" is held up against the other. The controversy over *Prout's hypothesis* in the nineteenth century may serve as an illustration.[14]

In 1815, William Prout postulated that all atoms are made up of atoms of hydrogen, the lightest element, as the fundamental building block. This implied that the atomic weights of all chemical elements would have to be exactly whole numbers (measured in hydrogen units); many chemists performed painstaking experiments to check, but their results did not confirm Prout's idea—the atomic weights of most elements deviated from the predicted values. We now know that Prout was essentially right, and we also understand why the many measurements of reputable chemists were correct but irrelevant: most naturally found elements are mixtures of various isotopes—substances with identical chemical properties but with atomic weights that differ by small multiples of hydrogen units. Before facts can be used to substantiate or refute a theory, they have to be understood in the light of other theories.

On the basis of the examples given, one may well argue that the *dividing line between facts and theories is not always sharp.* Furthermore, in addition to the blurring of the line between theories and "data," experiments are also subject to influences that may be irrational. After all, their interpretation always has to be guided by ideas based on previous thoughts and knowledge. To perform and evaluate experiments without preconceived ideas, Poincaré noted, "is impossible. Not only would it make all experiments barren, but that would be attempted which could not be done. Everyone carries in his mind his own conception of the world, of which he can not so easily rid himself."[15] So the genesis of what we regard as facts is, in many senses, adulterated.

Stability of Facts

What, then, of the oft-repeated assertion that the imaginative theories of science are subject to the ultimate test of having to be anchored in facts and experimentally verifiable evidence? If these facts are generated by experiments that are full of preconceived ideas, and if they depend on the very theories they are to corroborate, are we to conclude that science is nothing but a vast conspiracy or myth in which the theories are used to fabricate their own facts? Do the results of experiments have no more cognitive value than folklore? There are, alas, supposedly intelligent people who contend just that, but such a conclusion would be quite foolish.

Even though Römer's first determination of the velocity of light relied upon astronomical theory, this speed has since been measured in many different ways, all of which lead to consistent results. Rutherford's discovery of the tiny atomic nucleus, based though it was on theoretical calculations (and using incorrect theory at that), has been corroborated innumerable times by other experiments and used for a vast variety of successful predictions. The numerical value of the electric charge of electrons enters into many precisely verified predictions. Its special combination with Planck's constant and the speed of light, called the *fine structure constant,* which is "dimensionless"—the numerical value 1/137 does not depend on the units of measurement in which its separate ingredients are expressed—is now most precisely determined by experiments in condensed matter rather than particle physics. The masses and half-lives of elementary particles manifest themselves in high-energy experiments of different kinds, and the relativistic time-dilatation that accounts for the stretching out of the half-lives of unstable particles has been verified in a variety of different contexts. The fact-theory combination of astrophysics, too, is becoming increasingly sturdy and secure, though because some of its "facts" may yet turn out to be illusory, this is an area of physics that does not yet inspire as much confidence as others do.

While it is a great oversimplification to say that theories and laws are finally based on independently verifiable facts, it is nevertheless the case—and this needs emphasizing—that in physics and in most other areas of science the combination of laws and facts, theory-laden though many of the latter may be, has an enormous amount of stability. Just as the stability of a large building depends not simply on the firmness of its corner posts but is due in large measure to the ubiquitous cross-bracing in its walls and floors, so the structure of science is secured by the dense network of interdependencies that are established among its various parts. Pointing out, when necessary, that the evidence for certain specific parts of science may be weak is an important function of scientists and knowledgeable commentators, but it has little bearing on the safety of the edifice as a whole.

6

THE BIRTH AND DEATH OF THEORIES

AS the last chapter indicated, the line that separates the laws and theories of science from the factual evidence is not sharp. Nevertheless, we need to ask what is the relationship between the two? For that matter, what are the origins of theories and what determines their status? Do they emanate from these somewhat contaminated facts? Are the laws ultimately *proved* by them?

The Origins of Ideas and Theories

To begin with, it is important to make a clear distinction between the origin of a theory and its confirmation. The contention of some philosophers and sociologists of science notwithstanding, the germination of a law in the mind of a scientist—its psychological origin—has very little relation to the evidence on which the law rests.[1] That is why the professional writings of scientists tend to be so impersonal—to the despair of biographers and historians. Scientists screen out all the internal wrestling they went through and the doubts that had to be overcome before they present their theories in a shape they deem suitable for inspection and testing by the world. "The success of science as a shareable activity," the science historian Gerald Holton rightly observes, "is connected with the conscious downplaying of the private struggle."[2]

Scientists seldom arrive at a theory by poring over a mass of experimental data and following some systematic method of induction, an image of the "scientific method" that Karl Popper was instrumental in helping to demolish once and for all. We have abundant testimony from scientists themselves that the invention of a theory is almost always an act of imagination and a flash of inspiration. In fact, in many cases, the sudden insights have no relation at all to anything we would call evidence, especially since these intuitions are sometimes based more on subconscious thoughts than on rational reflection.

To find examples, we do not have to go as far afield as the discovery by Paracelsus of the curative power of mercury for syphilis, the result of his astrological belief in the significance of the historical circumstance that the planets Venus and Mercury were in opposition when he made his proposal. There is the well-known story of how Kekulé von Stradonitz intuited the stereochemical structure of the benzene ring from a vision he had, in a half-sleeping state, of a snake biting its own tail. Perhaps more to the point is Einstein's recollection, which you may remember from Chapter 4, of how the germ of the (special) theory of relativity came to him. As a boy, sitting in a streetcar on the way to school, he would imagine what the world would look like if he were riding along on a beam of light. It is not entirely clear whether, at the time he formulated his theory, Einstein actually knew about the Michelson-Morley experiment,[3] performed some eighteen years earlier, but this experiment did not play an important role in the thinking that led him to the theory of relativity. What is certain, though, is that the experiment is one of the many pieces of corroborating evidence and plays a crucial role in establishing the validity of the theory in the minds of students. Similarly, Hans Christian Oersted was, from his reading of Immanuel Kant, metaphysically convinced of the existence of a fundamental connection between electricity and magnetism before his discovery of the influence of an electric current on a compass needle.[4] The point is not that experimental data are unnecessary for generating a law or theory, but that sometimes many such data are needed and sometimes a flash of intuition occurs on the basis of just a few; at the psychological source of an idea, a great variety of influences, including social and political, may come into play as well. The discovery of the first law of thermodynamics—the conservation of energy—is an example worth looking at in some detail.

One of the most exciting issues in physics in the first half of the nineteenth century was the nature of heat and its relation to mechanical energy. The physiologist Julius Robert Mayer, often credited with the discovery that heat is equivalent to mechanical energy, came to his conclusion in a remarkably indirect manner. While serving as a ship's doctor in 1840 on an extended voyage to the East Indies, he observed that the color of the blood of European sailors was much brighter when in tropical climates than it was at home. He concluded that the body must use less oxygen from the blood to maintain its temperature in warm surroundings. Since, he generalized, the only source of animal heat was the oxidation of foodstuff via the blood, it must, therefore, be the total amount of chemical, mechanical, electric, magnetic, and heat energy that was conserved, and he used this reasoning to argue forcefully against the vitalism—according to which life required an irreducible and unique force, *vis vitae*—fashionable at the time. Mayer's conclusion, of course, was quite correct, but it was based on a somewhat confused theorizing that identified causation with energy, which he viewed as a unique, identifiable substance. Other German scientists, such as Hermann von Helmholtz, similarly arrived at the conservation law of energy from a physiological point of view and used it as a weapon against vitalism. On the other hand, the British contender for the credit of discovering the first law of thermodynamics, James Prescott Joule, approached the relation of heat to energy entirely from physics. Warming a bucket of water by rotating a paddlewheel immersed in it—in effect, directly producing heat by means of mechanical work—he measured both the amount of mechanical work done in the turning—or of the electrical work done when the wheel was turned by a motor—and the temperature rise of the water. From this famous experiment he determined the "mechanical equivalent of heat," that is, how many units of mechanical or electrical work were equivalent to one calorie of heat.

The history of the conservation law of energy is a convoluted story, full of priority disputes that echoed through many succeeding decades. The lesson to be learned is that this extremely important physical law was discovered along various different routes by scientists who did not always base their reasoning on ideas we still recognize as valid. The same is also true of the second law of thermodynamics, which postulates the constant increase of entropy in an isolated system—heat flows always from hot to cold. The earliest version of this fundamental law came from the French engineer Sadi Carnot, who still believed in the old-fashioned caloric theory that heat was a separate substance. Instances like this one make it quite clear that the origins of valid theories are ultimately irrelevant. Scientists are well aware of the fact that the first proposal of an important new insight is often based on flawed reasoning; this is why science textbooks are usually written in an ahistorical style, substituting an idealized and anachronistic description for the actual course of events that led to a discovery.

The theory of electromagnetism represents a large and enormously fruitful set of ideas originating from underlying concepts that were soon discarded and considered irrelevant. Maxwell based his equations, which combined the previously quite separate areas of electricity and magnetism, on a complicated mechanical model that he believed characterized reality (see Figure 4). He then used these equations largely for an explanation of the nature of light and demonstrated that they implied all the known laws of optics. It was Heinrich Hertz who deduced from them the existence of other electromagnetic waves in the air or in a vacuum and experimentally produced and detected such waves—with enormous technological consequences. And it was Hertz who announced at the beginning of his book, *Electric Waves*, that "Maxwell's theory is Maxwell's system of equations," thus severing the connection between the underlying model and the theory, as it remains to this day. No author of a modern textbook on electromagnetism feels the need even to mention the model on which Maxwell based his equations; there are now other ways, more in line with contemporary views of physics, of arriving at them intellectually. What ultimately justifies them, of course, are their many experimentally verified consequences.

There are times when the intellectual climate in which the scientist finds herself is conducive to productive new ideas in a certain area, and other times when it is hostile. Furthermore, scientists with different temperaments naturally react in different ways to their environment. Not only may individual scientists thus suggest divergent answers, but they may ask disparate questions. The very selection of what needs an explanation may be subject to influences that are not always scientific or even rational. From many directions, it seems, varieties of style enter the arena of scientific reasoning.[5]

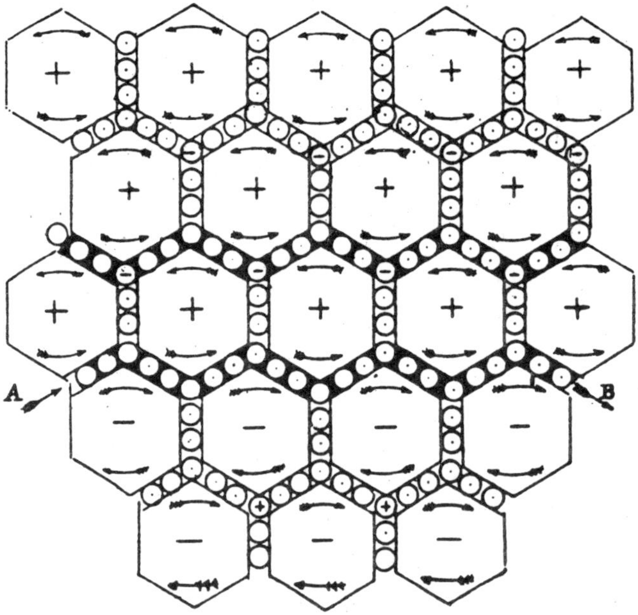

FIGURE 4 Maxwell's model of a magnetic field producing an electric current. (Reprinted from John Hendry, *James Clerk Maxwell and the Theory of the Electromagnetic Field* [1986], p. 172, with kind permission from Adam Hilger, Ltd., Redcliffe Way, Bristol BS1 6NS, U.K., and from the author.)

Some scientists are more comfortable conceiving of theories by unifying masses of data; others choose on the basis of the manifest beauty of a solution to a puzzle. For Paul Dirac, the motivation was always largely esthetic—few physicists would deny the appeal of the Dirac equation—and he was by no means alone in this. As the French mathematician Jacques Hadamard put it: "invention is choice, [and] this choice is imperatively governed by the sense of scientific beauty."[6] The validation of an idea or a theory, however, is entirely independent of its origin. The Dirac equation, combining quantum mechanics with the theory of relativity, is not regarded as a valid description of the behavior of electrons and other particles because of its much appreciated esthetic appeal but because it has explanatory and predictive power and is well corroborated by experiments. A law of Nature is not accepted simply because a particularly colorful vision led to its discovery or because the *Zeitgeist* made it desirable.

That is the crucial difference between science and other intellectual activities.

Every good scientist, of course, has a filter that screens out unsuitable ideas before they gel into a theory thrown open for testing—namely, his own knowledge of the laws of Nature as they have been established. "Our science makes terrific demands on the imagination," Richard Feynman said:

> The whole question of imagination in science is often misunderstood by people in other disciplines . . . [W]hatever we are *allowed* to imagine in science must be *consistent with everything we know:* the electric fields and waves we talk about are not just some happy thoughts which we are free to make as we wish, but ideas which must be consistent with all the laws of physics we know. We can't allow ourselves to seriously imagine things which are obviously in contradiction with the known laws of nature . . . One has to have imagination to think of something that has never been seen before, never been heard of before. At the same time the thoughts are restricted in a strait jacket, so to speak, limited by the conditions that come from our knowledge of the way nature really is.[7]

This constraint on the imagination keeps scientists from flying off into the Oz of science fiction; it also tends to make them relatively conservative, generally reluctant to rock the boat too much. There have been a few rare occasions on which great scientists have indeed allowed themselves "to seriously imagine things which are obviously in contradiction with the known laws of nature," and these radical new ideas have revolutionized our view of the world. Niels Bohr's model of the atom was such a "paradigm shift." The configuration he postulated—of a stable "solar system" of electrons circling the nucleus in certain specific orbits—seriously violated the known and well-established laws of electrodynamics. Similarly did Isaac Newton postulate laws of motion that violated "the known laws of nature," formulated by Aristotle and accepted for almost two thousand years. And while Einstein's special theory of relativity did not directly violate "the known laws of nature," it seriously contradicted our usual way of thinking about space and time, and to this day it offends the logical sensibilities of many nonscientists. Science is a conservative enterprise with a revolutionary escape clause, like a sturdy steam kettle with a well-functioning safety valve.

The same constraints on the imagination that exist for scientists also exist for mathematicians, but in their case, of course, these constraints originate from the the rigorous demands of logic and the whole body of mathematics rather than from the external world. But great mathematical insights leading to important new theorems often occur and, as is true for scientists, mathematicians sometimes have epiphanies that are not always traceable to rational cogitation. In both cases they appear as sudden solutions to puzzles or problems. Here is how Hadamard described it:

> One phenomenon is certain and I can vouch for its absolute certainty: the sudden and immediate appearance of a solution at the very moment of sudden awakening. On being very abruptly awakened by an external noise, a solution long searched for appeared to me at once without the slightest instant of reflection on my part—the fact was remarkable enough to have struck me unforgettably.[8]

And what are we to make of the great Indian mathematician Srinivasa Ramanujan, who arrived at the most astonishing results in the theory of numbers with no proofs at all? It took other mathematicians years to establish that many, but not all, of the complicated and elegant formulas he had written down without any derivation were correct. For much as other mathematicians admired his beautiful results—and may have been inclined to give them the benefit of the doubt because of their beauty—they could not accept his propositions without rigorous proofs, just as scientists insist upon experimental evidence for their theories, no matter how elegant they might be.

Since scientific theories often are based on flashes of insight and produced by a fertile imagination, and since, in addition, they are radically *underdetermined* by the supporting evidence, we should expect to find instances in which several different competing theories are proposed to account for the same group of phenomena, perhaps all equally effective and confirmed by experiments. So it happened in 1925 when Heisenberg proposed his "matrix mechanics" for the quantum theory and Schrödinger his "wave mechanics." Both theories accounted equally well for the quantum puzzles but were seemingly totally different. It was not long, however, before the two formulations were found to be mathematically equivalent to one another, and they are now seen simply as two versions of the same abstract quantum theory. "The contentious issue," as the philosopher Brian Ellis sees it, "is whether there are genuine, logically incompatible, theories which are empirically equivalent, in the *strong* sense that no evidence could possibly distinguish between them."[9] I know of no case in which two inequivalent theories of equal range were proposed for the same area,[10] which is surely remarkable in view of the fact, emphasized by many philosophers of science, that theories are never logically determined by observational data.

Fashions and Fads in Science

Given the murky, sometimes irrational psychological origin of theories, it can be no great surprise to find that theorizing in science is often subject to *fashions.* The existence of trends that catch us up in their currents more or less without our volition is often cited as the hallmark of nonscientific, "irrational" intellectual disciplines, and scientists sometimes sneer at scholars in the humanities for allowing themselves to be seduced by whatever idea happens to be in vogue at the moment. There is, however, no doubt that fashions exist, to various degrees, in science and mathematics, too, even if they do not play as dominant a role there as elsewhere. Fads can lead scientists—especially young ones, out to make their mark and to jump-start a career—to ask specific kinds of questions or tackle a certain class of problems, not necessarily because these problems are the most important or appear to be most amenable to solution with the tools of the time, but because they have been raised or are being discussed by "everybody," especially by the most prominent scientists in the field. It is hard to deny that such social pressures, deplorable as they may be from a purely rational point of view, do exist and exert an influence on science.

Examples of fads in physics are not difficult to find. The "bootstrap" method in particle physics, so prominent in Pickering's attack on the worth of modern physics described in Chapter 2, was clearly a fashion that came and went with bell-bottom trousers. Catastrophe theory was another "hot" idea that was supposed to solve, simply and elegantly, a great variety of puzzles bedeviling scientists in many specialties. It fell far short of its goals, as did fractal theory and chaos to a large extent. This is not to say that these pursuits were or are devoid of merit. What characterizes them as fads is the wave of avid fascination and sudden loss of interest among a multitude of scientists, which begins with the quick production of large numbers of publications yielding relatively few meaningful results and ends with the almost total disappearance of attention after their allotted "fifteen minutes of fame." If it is principally the discrepancy between the temporary supercharged activity among many researchers and the relative insignificance of the final outcome that distinguishes fashions from other rapid changes in scientific focus, the catch is that it is not always easy at the time to distinguish a voguish craze from an important advance, and *vice versa.*

The existence of trendy fads in science must not be confused with the very natural, healthy loss of research interest among most scientists—and to a lesser extent, among mathematicians—in areas that are regarded as overgrazed. No matter how old a field of research, and how well understood it is in general, there are always many specific, detailed questions that have not been satisfactorily answered; a scientist who spends time and effort to wring an extra drop from a cow that is milked dry will be rewarded neither by fame nor monetary support. A rapid change of focus, like the one that occurred in biology when it shifted its attention from macro to micro structures—from large animals and plants to cells and microorganisms—can devastate the old-fashioned practitioners, who suddenly feel disenfranchised. Not every change of direction is a veering of fashion, however; sometimes a new orientation is more in the nature of a "paradigm shift." The area that has been abandoned is not necessarily falsified by the change of focus, but the unanswered questions left in the old research program are no longer thought to be interesting.

How Theories are Tested

Even the ideas that pass the fine filter posited by Feynman are, of course, not always correct—many imaginative theorists have had ideas that were not obviously wrong but that proved to be incorrect when checked by observation. Others followed Bohr's and Einstein's lead in violating established laws and were found to violate the evidence as well; that some "crazy ideas" turned out to be right does not justify all unorthodox notions. We might wonder, then, what determines whether theories, old and new, are accepted or rejected.

The first thing to be said is that, just as many kinds of nonrational elements play a role in the psychological origins of a theory, so they may in individual scientists' instant rejection or easy acceptance of it; initial judgments, even by the the most astute, are sometimes found to be quite incorrect. Galileo's refusal to accept Kepler's laws of planetary motion seems quite inexplicable in view of the large part Copernicus's new astronomy played in his thinking—for which, indeed, he was persecuted by the Church—and the powerful support Kepler's laws gave to Copernican theory. This puzzling behavior probably had its origin in Galileo's strong esthetic preference for the circle and his abhorrence of the esthetic of "mannerism" in painting and scupture, with its fashionably elongated figures that stretched circles into ellipses.[11] For Galileo, Holton argues, Kepler was a *mannerist* thinker and therefore to be dismissed.

Ultimately, to be accepted, any proposed scientific law has to lead to verifiable consequences. As a matter of general philosophy, the positivists insisted that propositions which could not be tested by observation had no meaning. This requirement, however, is much too stringent, especially since the word *meaning* can signify many different things. Science, particularly the abstract discipline of physics, is full of concepts that are far removed from experimental significance and of meaningful statements that, in and of themselves, have no observational consequences. Nevertheless, there is no question that a law would be unacceptable if none of its implications were subject to experimental testing. A very general law, as I have noted earlier, may sometimes spawn local laws which are more likely to lead to directly verifiable propositions—predictions of the results either of future observations or of an appropriate analysis, so far unperformed, of past experiments (these might be called "postdictions," in the sense that they agree with facts already known).

Generally speaking, scientists put a much higher premium on predictions than on postdictions. The reason is largely psychological. There is a feeling that with sufficient jiggling of adjustable parameters, clever theorists can always come up with a scheme that fits a known set of data; it is in the predictions that even the smartest have to risk being wrong; this is why the greater the number of significant new predictions an announced theory leads to, the more highly it is valued. Predictions also have the virtue of stimulating experiments that might otherwise not have been performed; when asked a new set of questions, Nature may offer up a new set of very revealing answers. Of all the potential values of a new theory, this is the most important. The planet Neptune was discovered in the middle of the nineteenth century after J. C. Adams and U. J. J. Leverrier independently analyzed irregularities in the motion of Uranus on the basis of Newton's gravitational theory and predicted the position of the unknown planet that caused them. Their work was regarded as a brilliant corroboration of Newton when J. C. Galle and H. d'Arrest followed their prediction and observed the planet.

It seems obvious that the more observations and experimental results agree with the implications of a theory, appropriately interpreted, the higher is its degree of confirmation. However, there is no way of assigning a numerical value, such as a probability of the correctness of the theory, to that degree of confirmation. All attempts at doing so on the basis of a systematic inductive method have failed, and it was Karl Popper's great contribution to emphasize that ultimately what matters in giving meaning to a scientific law is its *falsifiablity* rather than its verifiability. This is true simply because laws have the form of a general proposition: such and such will always be found to be the case; or, such and such can never happen. Energy is always conserved; perpetual motion of the second kind is impossible—energy cannot be extracted from the sea by simply cooling it down. General statements like these can never be completely confirmed, because it would take an infinite number of attempts to do so. They can, however, be disproved by a single negative observation.

There is another reason, as well, why falsifiability rather than verifiability renders a statement scientifically meaningful—a verification of almost anything can be easily accomplished by appropriate selection of the evidence. Even the observational confirmation of a prediction may be based on a fluke: the calculations by Adams and Leverrier determining where to look for the unknown planet that disturbed the orbit of Uranus were based on faulty assumptions, and Neptune was found largely by luck.[12] "What makes the criterion of falsifiability so powerful is this," Ernest Gellner trenchantly observes,

> if you insist that a believer specifies the conditions in which his faith would cease to be true, you implicitly force him to conceive a world in which his faith is *sub judice,* at the mercy of some "fact" or other. But this is precisely what faiths, total outlooks, systematically avoid or evade . . . they have little to fear from a requirement that they be "verifiable": generally speaking, they pervade the world they create so completely that verifications abound—here a verification, there a verification, everywhere a verification.[13]

One of the most telling arguments against psychoanalysis as a science is that its system can easily produce plausible explanations of symptoms or dreams of any kind, but there appears to be no way to show that the explanation is wrong. If an explanation works for one person and fails for another, a reason can always be dredged up from the subconscious; the same explanation may be used to account for a symptom and its opposite, whichever is most useful or appears most plausible to a particular expert. Such theories thus have no real explanatory power; it is their non-falsifiability that accounts for their lack of scientific meaning, not their failure of verifiability. *There is no scientific proof; there is only disproof.*

The scientific importance of falsifiability accounts for the stress scientists put on predictions as necessary concomitants of explanations. So long as understanding a process yields no more than an explanation of facts or relations already known, it is safe and cannot be falsified; only when it leads to predictions does it become risky. That is why the fields of human thought in which explanation does not lead to predictions—even potentially—are full of unresolvable controversies, for the only reason to reject an explanation is that it is unconvincing.

Plausible and enticing as the simple and strict criterion of falsifiability may be, it has been subjected to forceful criticism by science philosophers such as Imre Lakatos. For one thing, it is usually possible to construct an *ad hoc* modification of a theory to account for an observational discrepancy. "*Some of the most important research programmes in the history of science were grafted onto older programmes with which they were blatantly inconsistent.* For instance, Copernican astronomy was 'grafted' onto Aristotelian physics, Bohr's programme onto Maxwell's."[14] For another, the test of any theory must always rely on some demarcation that gauges the outcome of an observation, *other things being equal.* In most instances, scientists have to make a judgment about what is responsible for a specific aberration. That the orbital perihelion of the planet Mercury deviated from its Newtonian prediction had been known to astronomers for eighty-five years, and yet it was not regarded as a disproof of Newton's gravitation theory. For all anyone knew, it was due to the perturbing influence of some as yet undiscovered planet or some other unknown cause. Only after Einstein's prediction of such a deviation on the basis of his general theory of relativity was it considered a genuine anomaly in the Newtonian theory, and since its observed numerical value seemed to agree with that prediction, Einstein had triumphed over Newton.

> "Falsification" in the sense of naive falsificationism (corroborated counterevidence) is not a *sufficient* condition for eliminating a specific theory: in spite of hundreds of known anomalies we do not regard it as falsified (that is, eliminated) until we have a better one.[15]

"There is no falsification before the emergence of a better theory,"[16] concludes Lakatos, and he proposes to replace "naive falsificationism" by "methodological falsificationism," which *"uses our most successful theories as extensions of our senses, . . . demarkat*[*ing*] *the theory under test from unproblematic background knowledge."*[17]

In other words, whether an observed discrepancy is regarded as a falsification of a theory depends on the experimental and theoretical context. In many instances, an experimental result that appears to disagree with a theory is regarded as erroneous. As noted earlier, when in 1929 some experimental observations strongly suggested that beta decay violated the law of conservation of parity, physicists' confidence in this principle was so strong that the data were ignored and buried, to be disinterred only twenty-seven years later, when the parity conservation law was finally found to be broken in other instances of a similar kind. Thus, in this case, falsification was swept under the rug to the detriment of progress. Nevertheless, there are more than a few statements by Einstein and Dirac asserting that they would not necessarily consider their theories falsified by a purported experimental result that disagreed with them. A steadfast belief in their theories even in the face of adverse evidence, if not carried too far, is often a virtue in theorists. The absence of observed phases of

Venus, for instance, presented a problem for Copernicus's revolutionary theory of the motion of planets. It was not until some fifty years later that Galileo could see these phases through his new telescope, and he praised Copernicus for sticking with his theory in spite of the hitherto unexplained puzzle.[18]

The falsification criterion therefore has inherent limitations; moreover, we must admit that "confirmation" also plays a significant role, even if this role cannot be quantified and must never be confused with "proof." There can be no question that when a new observation agrees with a prediction, it constitutes potent support for the validity of the theory to be tested and has a highly persuasive effect on scientists. When Julian Schwinger announced the result of his calculation of the anomalous magnetic moment of the electron, which agreed with the measured value to better than a few parts in 100,000, it was regarded as a triumph for his newly developed renormalization of quantum electrodynamics. The experimental discovery of the Ω^- particle was taken as glorious confirmation of Murray Gell-Mann's theory of the "eightfold way." John Polkinghorne is surely right:

> When Rubbia and his friends rejoiced in the outcome of the UA1 experiment and believed that they had discovered the W and the Z bosons, on a Popperian account their happiness was misplaced. Truth is always unknowable, the only certain knowledge is that of error. On that view, what would really have justified a party would have been the failure to discover the predicted W and Z signals! There is something very awry in such an account.[19]

The psychological point is well taken, especially when it applies to a young theory with few successes. However, it is nonetheless also true that there are many theorists who would indeed rejoice in the news of an experiment that refuted an apparently well-established theory and who would be bored with a result that merely tended to confirm it. The standard reaction of many physicists is to say that such a refutation teaches us *new physics*. For years now, large-scale experiments set up to measure the flux of neutrinos reaching us from the sun have found substantially fewer of them than expected on the basis of what we think we understand about their production by thermonuclear processes. It would excite us if all the convoluted attempts to account for the "missing neutrinos" from the sun turned out to be futile, for we would have to conclude there is something wrong with what is regarded as secure knowledge, either about neutrinos or the sun. When the antiproton was discovered in Berkeley at the Bevatron, which had been constructed specifically for that purpose, it sent a wave of ennui through the physics community. Not that its discovery was unimportant, but on the basis of Dirac's theory, everybody expected it. Had it not shown up, on the other hand, it would have been electrifying news, because that would have necessitated a complete restructuring of relativistic particle theory. As Max Planck once remarked in a lecture,

> A living and flourishing theory does not avoid its anomalies but searches them out, for the stimulus to further development comes from contradictions, not from confirmations.[20]

The point to keep in mind is that although falsifiability, rather than verifiability, is the most important *criterion* in determining whether a theory is scientifically meaningful, its usefulness for the greater task of building confidence in a theory is limited. A theory is accepted not simply because it has withstood many attempts at falsification, the need for such tests notwithstanding, but because it leads to predictions that are experimentally verified. After all, the purpose of a theory is to be productive and not just to fail to be wrong. "Only in the philosopher's never-never land," comments the philosopher Larry Laudan, "is it rational to espouse a doctrine simply because that doctrine has not been conclusively refuted."[21]

When deciding whether to throw over an old, falsified theory in favor of one that explains what the previous one could not, scientists are faced with still another question. Sometimes the new theory makes no predictions or postdictions about certain phenomena encompassed by the old. A case in point is Newton's theory of the planetary orbits, based on his laws of motion and gravity, which supplanted a prior Cartesian theory. The latter purported to explain, among other things, the fact that all the planets rotate about the sun in the same direction, whereas Newton's theory is silent on this point. We now look to the contingencies of history, to the way in which the solar system happened to be formed, for an explanation of this unidirectionality, as well as for explanations of the coplanarity of the orbits, the specific sizes of the planets, and their distances from the sun. A new theory not only brings with it new predictions contravening its predecessors—these contraventions are the tools by which the old theory can be falsified—but it also often changes the set of questions regarded as worth answering. This change in focus is the essence of a paradigm shift.

Crucial Experiments

Are there *crucial experiments* that test two conflicting hypotheses, establishing one and demolishing the other? Though such definitive trials are often cited in textbooks, there are science philosophers who deny their existence. Pierre Duhem persuasively argued against them because they can at best decide against a hypothesis in the context of a larger theory; the same hypothesis within another context may not be contradicted by the experiment. He cites[22] Léon Foucault's famous experiment pitting Newton's corpuscular theory of light against Huygens's wave theory. The former accounted for refraction of rays at the surface of water by assuming that light moves faster in water than in air, whereas according to Huygens light moved more slowly in water. Foucault's experiment led to the definite conclusion that the speed of light in air exceeded that in water, which was interpreted as a crucial blow against the corpuscular theory in favor of oscillations. Duhem, however, argues that while the outcome of Foucault's test could be more easily explained by waves than by particles, there might well be a different corpuscular theory of light that would survive it. And indeed, as Louis de Broglie points out in his introduction to a later edition of Duhem's book, at the very time of his writing (in the year 1905) and unbeknownst to Duhem, Einstein was introducing the quantum theory of light, which did just that. Duhem clearly made a valid point.

Nevertheless, another experiment in the same context should also be mentioned. The great French mathematician and physicist Siméon Denis Poisson, a prominent opponent of the wave theory of light, showed that one of its absurd consequences was a bright spot at the center of the shadow of a circular disk. When Dominique François Arago demonstrated that light indeed produced such a bright spot (see Figure 5),[23] he had conducted a crucial experiment in favor of the wave theory, one that convinced even Poisson. While the quan-

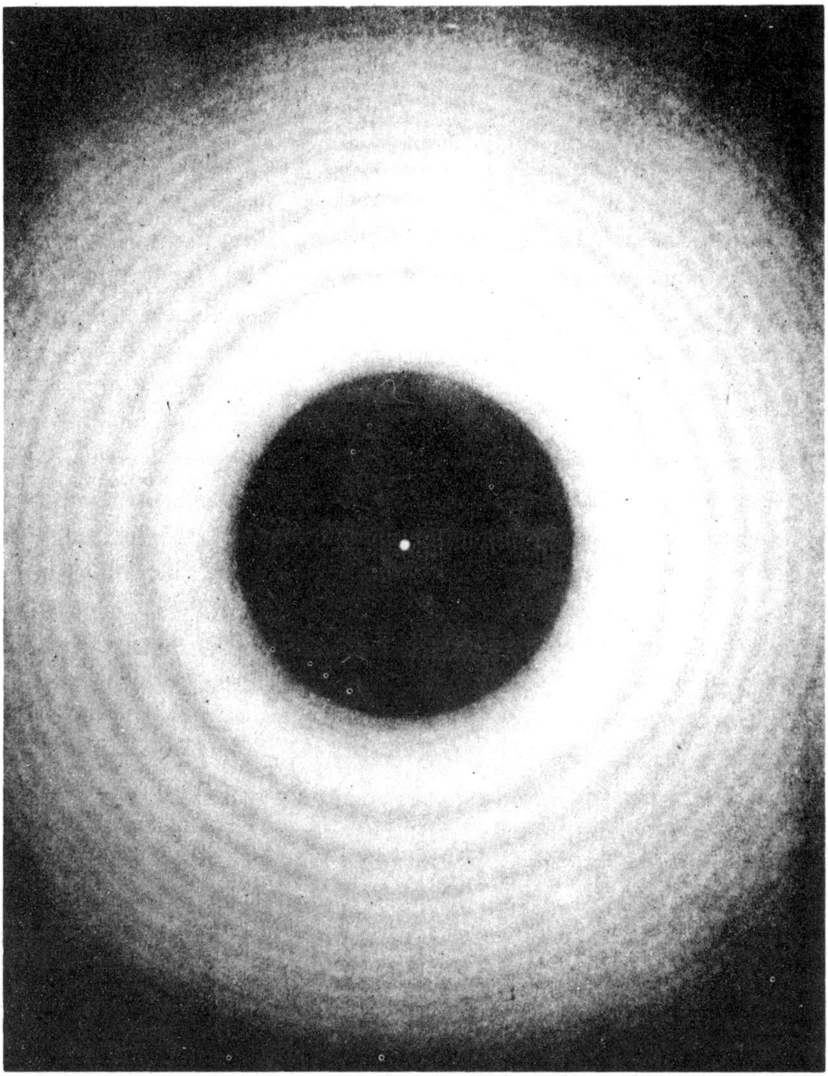

FIGURE 5 Diffraction of light by a circular disc. Note the bright spot in the center. (Reprinted from M. V. Klein and T. E. Furtak, *Optics* [1986], p. 440, with kind permission from John Wiley & Sons, Inc., 605 Third Avenue, New York, NY 10158-0012.)

tum theory shows this experiment is not necessarily fatal to all particle theories of light, no theory that denies waves can survive it. (The strange quantum situation produced by this paradox will be discussed further in Chapters 9 and 10.)

There are two kinds of experimenters, those who like designing laboratory tests to corroborate theories, and those who prefer trying to confute them. When a theory is new and on probation, its confirmation brings glory; when it is well established, fame attends the refutation. Of course, once a theory has been accepted after many trials, the chances of producing a negative result are very slim, and the choice confronting the experimenter is that facing any gambler—do you go for low-risk, low-payoff confirmation or high-risk, high-payoff refutation? The decision is mostly a matter of personality. If this makes scientists resemble human beings, subject to human emotions—rather than "demigods on stilts," in a phrase Einstein used in another context—it is because that is what they are.

Some twenty-five years ago, a flurry of theoretical speculation arose, good arguments against their existence notwithstanding, that there might be particles, named *tachyons,* traveling faster than light. While most physicists were very skeptical, a few intrepid experimenters began to search for these strange objects, which would speed up as they radiated light and lost energy. In spite of years of futile hunting, they could find nothing; if they had, however, the implications for physics would have been immense. The courage of experimenters who take a big risk of failing is very healthy for science and can only be applauded.

In addition to data that falsify or corroborate a theory directly, there are also facts that serve to increase or decrease our confidence in it more indirectly. As Brian Ellis points out, "the set of possible empirical discoveries relevant to [a theory's] truth or falsity" is open: "There can be evidence for a theory which is in no way entailed by it, and evidence against a theory which does not contradict it."[24] The finding that after a substance was burned its ashes weighed more than the original certainly counted against the phlogiston theory—according to which phlogiston escaped from the substance in the process of burning—even though the theory could accommodate this result by assuming phlogiston had negative weight. When a "phenomenological" local theory, based on a general theory in an intuitive and approximate manner, is corroborated, it is often used as testimony for the general theory, although there is no strict logical connection. It is clear, therefore, that there is more to what counts as evidence for or against theories than strict falsification or corroboration.[25]

Is There a Scientific Method?

All that I have described above—the way theories depend on facts, how theories are generated, and how they are confirmed or confuted—is sometimes encapsulated in the phrase *the scientific method.* Tomes are written defining and refining it, controversies surround it, and its very value is questioned. Textbooks and college courses of psychology and sociology begin by explaining to students what this method is and then exhorting them to follow it: you must rely on evidence and on evidence only; follow a strict procedure and protocol in your laboratory and do not discard data that disagree with your theory or conjecture; always remain objective and do not allow your passionate attachment to a preconceived idea to color your judgment; remain dispassionate and disinterested. Valuable as such advice undoubtedly is, it is futile to attempt to restrain the scientific method by means of a straightjacket and insist that rigid rules be followed always and exclusively. In his book, *Against Method,* Paul Feyerabend argues provocativeley against all such strictures and, despite a number of objectionable contentions, makes some important and wise observations. It may be overstating the case to say, as he does, that the "only principle that does not inhibit progress is: *anything goes*"[26]—but Percy Bridgman, a great physicist with a philosophical bent, pithily defined the essence of the scientific method in a similiar vein: "to use your noodle, and no holds barred."

Feyerabend also maintains that "*the events, procedures and results that constitute the sciences have no common structure;* there are no elements that occur in every scientific investigation but are missing elsewhere."[27] While this claim, too, is exaggerated, I want to emphasize that the present book confines itself largely to the physical sciences, and much of what I am saying does not necessarily apply to the disciplines whose structure is less general and abstract. In fact, many of the problems discussed here do not arise in other scientific areas, which have their own kinds of difficulties to contend with. I agree with Feyerabend that the methods of the sciences differ among one another in many respects and certainly have some components in common with other activities of the mind.

There can, however, be no question that its general approach and methodology, inadvisable as it may be to define it too narrowly, separates modern science from most other intellectual endeavors, and especially from the ways in which explanations of natural phenomena have been generated in the past (and still are by the vast majority of humanity). The one general theme that runs through all the sciences is that they rely on *evidence accessible to others,* or as the physicist John Ziman expresses it, on *consensible* knowledge.[28] How that goal is to be achieved cannot be prescribed in any detail, but the attitude embodied in modern science is by no means what one could call "natural," as Feyerabend's or Bridgman's definition might lead us to believe. Its most important characteristic is to rely neither on authority nor on individual revelation or intuition. That is not to say that scientists and mathematicians never take the word of an authoritative and highly regarded colleague on faith. They often do; no one can go through the details of every mathematical proof or repeat every experiment. Certainly researchers rely on the word of others in the community, but that reliance is not on authority *qua* authority—a statement is not accepted automatically simply because it is made by a person in command or written in a sacred book; recourse to one's own reasoning and senses or those of one's peers is always *implied.* Here is where the issue of accreditation of experts arises. "Nobody knows more than a tiny fraction of science well enough to judge its validity and value at first hand," argues Michael Polányi.

> For the rest he has to rely on views accepted at second hand on the authority of a community of people accredited as scientists. But this accrediting depends in its turn on a complex organization. For each member of the community can judge at first hand only a small number of his fellow members, and yet each is accredited by all. What happens is that each recognizes as scientists a number of others by whom he is recognized as such in turn.[29]

As for individual intuition and insight, of course scientists and mathematicians make essential use of them in arriving at new ideas, but that is never the end of the matter. Ultimately, the test of the idea is empirical and *public.*

It is in part because psychoanalysis and other theories of psychology do not rest on publicly accessible evidence but rely on the authority of masters that they have been attacked for not being scientific. The writings of psychoanalysts, in fact, often consist of conflicting exegeses of the works of the founder. On the other hand, the rise of behaviorist psychology was a reaction, in the name of science, against too much reliance on introspection, on evidence that is not public.

The Devonian controversy in the early nineteenth century demonstrates why it is not a good idea to leave the solution of a scientific problem to a relatively confined "in group." The dispute concerned the correct identification of certain geological strata in the county of Devon in England. Determining the age and origin of these formations was important because some of the deep layers contained fossil records of the early history of life on earth and, in addition, coal and minerals of great industrial value. Both the uniqueness of the Devonian findings and their interpretation in terms of disputed geological history were hotly debated scientific questions. These questions could not be resolved within the original circle of combatants, consisting almost entirely of members of the Geological Society of London, but once exposed to the air of a wider context and a larger group of participants from all over Europe and North America, the controversy was satisfactorily settled.[30]

When Theories are Superseded

What happens to superseded theories? In general, two destinies are possible. In some cases, the old theory, falsified by specific repeated experiments, is replaced by a new one, as in the case of Aristotle's theory of motion. Newton's laws—that an object remains in a state of rest or uniform motion unless acted upon by a force, in which case the object experiences an acceleration proportional to that force—replaced Aristotle's notion that a force was necessary to keep an object in motion. And this replacement occurred in spite of the fact that Aristotle's idea corresponds very much to our everyday experience, which makes it hard to eradicate even today; this was surely a "paradigm shift" in Kuhn's sense.

Let me cite two other instances of falsified concepts that were simply discarded and replaced by new ones. For most of the eighteenth century, chemists believed in the phlogiston theory of combustion, mentioned earlier. They thought that when matter burned or a metal corroded, the cause was the extraction of a substance, phlogiston. When the burned or corroded residue was found to weigh more rather than less than the original, *ad hoc* assumptions—that phlogiston was an "immaterial principle" rather than an actual material substance, that in the course of burning phlogiston did not simply escape but was replaced by something else, or that it actually had negative weight—were introduced to "save the phenomena." Finally, though, the detailed experiments of Antoine Lavoisier, on which the discovery of the element oxygen was based, demolished the phlogiston theory and led to its replacement by the concept of oxidation.

Another abandoned idea of the eighteenth and early nineteenth centuries was the caloric theory, according to which heat was an indestructible, fluid-like substance that filled the interstices of all solids, liquids, and gases and had the intrinsic property of flowing from hot to cold, just as water flows downhill. After protracted struggle, it was replaced by the kinetic theory—heat consists of the chaotic or vibratory motion of the molecules making up matter. It was the observations of Benjamin Thompson (an American loyalist who later became Count Rumford) that the boring of cannon barrels produced prodigious amounts of heat, and those of Humphry Davy, that pieces of ice could be melted by rubbing them against each other, which provided the crucial evidence against the caloric theory.

In other cases, the old theory was not completely replaced and discarded, but its realm of validity was reduced. Here the theory of relativity and quantum mechanics are the prime examples. Einstein's theories of relativity contradict or modify Newton's theories—the special his equations of motion, and the general his law of gravitation. In contrast to the caloric and the phlogiston theories, however, Newton's theories were not dropped but are still considered valid as "limiting cases"—limited to objects moving slowly in comparison with the speed of light and to objects whose masses are not too large. As a result, Newtonian mechanics is perfectly adequate and can be used quite successfully for most everyday purposes, even for the calculation of the trajectories of rockets sent to the Moon. No chemist or biologist needs to take Einstein's theories of relativity into account.

The situation in the case of the quantum theory, which was much more revolutionary than relativity, is somewhat similar. Here, the relevant scale is determined by the size of Planck's constant, though statements of exactly when the "classical limit" is valid are more complicated. Nevertheless, again, even though all matter is composed of constituents that are subject to the laws of quantum mechanics and quantum field theory, we are perfectly free to use "classical physics" without noticeable error for the calculation of the trajectories of billiard balls and planets, as well as for the construction of electrical power plants. In other words, classical physics, though in principle no longer valid, is now embedded in both relativity and the quantum theory in such a way that its region of validity is restricted to the scale of everyday living; with few exceptions, its breakdown is noticeable only on the scale of the very large—astrophysics and cosmology—and the very small—molecules, atoms, and their nuclei.[31]

The state of physical science is now at a point that makes it unlikely we will ever again see a basic general theory superseded in the sense of being totally abandoned, except perhaps a theory of history such as cosmogony. No doubt many of our present general theories will have to be revised, but they are likely to remain as limiting cases or phenomenological constructs enclosed within future theories. The present theories of physics and chemistry account for too many details with great accuracy to be given up altogether. Even if a future sub-micro theory should supersede it, the quantum theory is likely to remain as an enormously useful framework, valid in its domain of applicability. Some of our local theories, on the other hand, might have to go by the board, just as other sciences that are at a less developed stage may well see some of their present theories totally abandoned.

In the course of my description of the role and structure of the general theories of the modern physical sciences, it has become abundantly clear that mathematics plays an enormously important part in the manner in which these theories are formulated and even sometimes in the way they originate. Let us, therefore, now turn to a more detailed examination of the nature of mathematics and learn why it is indispensable for the pursuit of science.

7

THE POWER OF MATHEMATICS

AFTER his release in 1832 from nine months in jail for republican agitation against King Louis-Philippe, Evariste Galois, an impetuous young French mathematician, died in a duel at the age of 20. The night before the fatal fight, in which he expected to perish, he had hastily written down the fruits of his latest research on the solvability of algebraic equations, results which became the foundation of what is now called the *theory of groups,* an esoteric and abstract branch of algebra. More than a century later, a fundamental particle of Nature was discovered whose existence had been predicted by the American physicist Murray Gell-Mann on the basis of that esoteric theory. Group theory now forms one of the cornerstones of our understanding of Nature at its most basic level and is used as an important explanatory tool in many other areas of physics, from materials science to nuclear physics. Indeed, all of chemistry rests on the periodic table of the elements, whose structure, in turn, is founded in large part on group theory.

Mathematics can be seen everywhere in science—in biology, chemistry, psychology, and most pervasively in physics. "The language of Nature is mathematics," Galileo declared, and the more abstract and general the discipline, the more pervasive is its use, with physics in the lead. Psychology and sociology, though dominated by the use of statistics, are still relatively unmathematical. Biology, on the other hand, is now becoming increasingly abstract, and the more structure it acquires at the molecular level, the more it can be expected to utilize mathematical tools as well. It would, however, be a mistake to believe that mathematics, whose use scientists have to learn but nonscientists fear and shun, serves only as a conventional and convenient language or as a means for generating numbers that can be checked by experiments. Mathematics is an enormously versatile and powerful instrument of thought, and physicists employ its results as much as its language. In this chapter I want first to examine both its historical and functional roles and then to ask where the power of mathematics originates. Why is it so effective?

The Historical Importance of Mathematics in Physics

Since Galileo, the theories of physics have been formulated in the language of mathematics—the terms "mathematical physics" and "theoretical physics" were, until this century, used interchangeably. One of the greatest of modern mathematical physicists, Henri Poincaré, likened physics to a large library in which the experimenters furnish the books and the mathematical physicists the index, without which the collection would be inaccessible and useless. "By showing the librarian the gaps in his collections," he observes in a comment that touches a nerve in today's world, "it will enable him to make judicious use of his funds; which is all the more important because these funds are entirely inadequate."[1] Accepting this simile, we should also note the remarkable fact that the nature of the index determines most of the organization of the library, and even the existence of many of the books. Both the language and the content of mathematics continue to have a powerful influence on our ideas of the structure of Nature; the results obtained by mathematicians, with no thought to applications, are almost always found to be both useful and—in a phrase of Eugene Wigner's—"unreasonably effective" in the physical sciences. Conversely, many important mathematical ideas originated from the needs of physics, which is why, from the seventeenth through the nineteenth century, most of the great mathematicians were also physicists.

The greatest of the combined physicists-mathematicians, Isaac Newton, invented the calculus, which in the course of the next two centuries grew into the large branch of *analysis* in mathematics, for the specific purpose of physical application—both in the formulation of his equations of motion and for the law of gravitation. Other instances in which advances in physics motivated progress in mathematics are easy to find. The need to solve Newton's equations of motion of objects in mechanics greatly stimulated progress in the area of differential equations, and the field of partial differential equations would not be as highly developed as it is without Maxwell's theory of electromagnetism, the mechanics of elastic extended bodies, or the theory of sound.

Here is an example of a discipline of mathematics arising from attempts to solve a specific problem in physics. When it became apparent how complicated the Newtonian theory of the solar system really was—each planet is gravitationally attracted not only by the sun but by all the other planets as well—the urgent need arose to ascertain mathematically whether the planetary system was stable or whether it might be destined to collapse. To prove the stability of the system, Joseph Louis Lagrange devised a method which engendered a new, productive branch of mathematics that came to be called *perturbation theory,* whose use now pervades all applications of theoretical physics. Many other great mathematicians found their imagination stimulated by working and contributing to science. The "prince of mathematicians," Karl Friedrich Gauss, contributed to mechanics, acoustics, optics, magnetism, and crystallography, and for many years he was director of the Göttingen observatory. In fact, astronomy is the only area in which he worked virtually his entire life. We might name as well Leonhard Euler, Jean d'Alembert, Pierre Simon de Laplace, Adrien Marie Legendre, Karl Gustav Jacobi, William Rowan Hamilton, and Henri Poincaré.

Not surprisingly, it was a mathematician, Laplace, who made the famous remark, which echoed throughout the nineteenth century, that sums up the view of the world engendered by theoretical physics:

> Given for one instant an intelligence which could comprehend all the forces by which nature is animated and the respective situations of the beings who compose it—an intelligence sufficiently vast to submit these data to analysis—it would embrace in the same formula the movements of the greatest bodies and those of the lightest atom; for it, nothing would be uncertain and the future, as the past, would be present to its eyes.[2]

Hailed by some, reviled by others, for a hundred years this pronouncement represented the grandiose image of deterministic physics until its impact was mitigated both by quantum mechanics and by the work of Poincaré. Though the determinism expressed by Laplace was mathematically correct as far as classical physics was concerned, almost all physical systems quickly become what we now call *chaotic.* In a chaotic system, the "sensitivity to initial conditions" makes prediction useless, in effect, because the smallest error in specifying the initial state is quickly magnified so as to render any forecast uncertain. (This is sometimes referred to as the "butterfly effect," after a colorful hypothetical example: a butterfly batting its wings on the Amazon might alter the course of a tornado in Wisconsin.)

While Laplace's pronouncement, based on the Newtonian formulation of the laws of physics by means of differential equations, stands as a synecdoche for the achievement of physics by means of mathematics, it also tacitly points to the limitations of this explanatory scheme. In order to yield predictions, the differential equations have to be supplemented by initial conditions, which are not supplied by physics or mathematics but by history. The course of the universe therefore retains an element of contingency, even in the most rigorous scientific description.[3]

The mathematical physics of this century differs from the earlier variety primarily because there are now many more areas of mathematics that are relevant to physics. From Newton's time until the end of the nineteenth century, only elementary algebra, Euclidean geometry, and classical analyis were used in physics, but Einstein formulated his general theory of relativity in the language of differential geometry, a powerful combination of analysis and geometry developed by Gauss and Riemann. The advent of quantum mechanics introduced into physics a heavy use of functional analysis—a vast generalization of analysis—and of abstract algebra; the development of the quantum theory of fields added measure theory (an outgrowth of the integral calculus), abstract probability theory, and, most recently, the theory of knots. With the addition of all these much more recondite areas of mathematics, what is now called "mathematical physics" became differentiated from "theoretical physics" and a schism began to form between the two. While every area of physics has practitioners on both sides of the divide and they all make extensive use of mathematics, those who regard themselves as mathematical physicists tend to pay more detailed attention to purely mathematical questions (which the others ignore) that arise within physical theories. In the following I will give some examples of the differing foci of these groups.

The Use and Purpose of Mathematical Physics

Our theories of fundamental particles are all based on quantum field theory, a fertile ground for problems in mathematical physics where the use of mathematics is twofold. In the first place, abstract areas of mathematics are extensively utilized for the formulation of theories—the theory of groups, embodying a mathematization of symmetry requirements, is one branch widely and powerfully employed, the latest formulation of differential geometry another. Second, rigorous mathematics is used to answer the question whether a given theory, expressed in the form of equations, makes mathematical sense. Do the postulated equations have solutions? Have they more than one solution? What are the properties of these solutions? How can the solutions be constructed for actual numerical calculations that can then be compared with experimental data? Sometimes, as in the case of string field theory, such investigations lead to deep new mathematical developments, or even to the construction of whole new areas of mathematics. To find a consistent theory that combines the quantum field theory of elementary particles with the force of gravity is one of the greatest challenges now faced by mathematical physics, and fresh mathematical concepts are being proposed to meet it. Some mathematical physicists even hope to be able to prove eventually that there can be only one completely consistent theory encompassing all the fundamental forces of Nature—a "Theory of Everything"—thereby proving, in Einstein's phrase, that "God had no choice in how He constructed the universe" because any alternative would be mathematically self-contradictory and, thus, literally unthinkable.

In quantum electrodynamics (QED), for instance, we are confronted with an unsolved puzzle: on the one hand, the theory has produced unprecedented precision in its agreement between experimental data and calculations based on well-defined approximation schemes for solving the complicated relativistic field equations of QED, which govern the interactions of electrons with photons; on the other hand, all attempts at actually proving that the equations, in fact, have solutions have failed for some forty years. If a proof that the equations of QED have "no solutions" should be forthcoming (and there are some indications that that may be the case), it will be the task of mathematical physics to lead us to an understanding of what it is, if it is not a "solution" of the equations, that has been calculated and that agrees with Nature so miraculously. The concept of what constitutes a solution of such equations will have to be appropriately redefined.

Such examples are compelling us to ask what the purpose of mathematical physics really is. Let me not beat about the bush: there are physicists who are hostile, or at least antipathetic, to the use of too much mathematics in science. Here I am not referring just to the lunatic fringe—men like the German physicists Philipp Lenard and Johannes Stark, Nobel Prize winners who in the 1920s attacked Einstein and his theory of relativity as an outgrowth of the Jewish mind—but scientists, both theoretical and experimental, who are unsympathetic to abstract formulations on much more respectable grounds. Michael Faraday in his early years is a prime example. Many physicists regard an explanation of an observed phenomenon that is based simply on a solution to an equation as unsatisfactory; looking for an intuitive, physical explication, they disdain theorists who cannot provide it.

A question that arises even in teaching a standard graduate course in electricity and magnetism at a university exemplifies one of the objections to an approach that is too mathematical: when dealing with a given system of partial differential equations, like the Maxwell equations with certain specified boundary conditions, which standard procedures appear to solve, what is the point of trying to determine if it actually has a solution? Some physics teachers take the attitude that such a mathematical question is pointless, because, after all, we know there is a solution: just look at the world. But this view misses the point that we can never be sure a theory, formulated as a set of equations, really describes Nature. If we find that a given system of equations does not have solutions of the kind we expect on the basis of experimental evidence, it does not follow that there is something wrong with Nature, but there is something wrong with the theory.

Relativistic quantum field theory and the quantum theory of many-particle systems made this question acute. In the case of the former, as mentioned earlier, we may have to reformulate what is meant by a solution of the equations of QED. In the case of many-particle systems, it turned out in the 1960s that nonrelativistic quantum mechanics was adequate to the task of proving the "stability of matter." If some physicists breathed a sigh of relief, it was not because they were afraid that matter might not be stable, but because they feared that quantum mechanics might fail the test.

Too Much Mathematics in Physics?

Too much mathematization of physics, runs another argument, is *sterile*. What is the point of rigorous mathematical studies of theories? Almost invariably we have to make approximations in order to be able to get experimentally verifiable predictions from them anyway, and that is where the meat of physics is. Furthermore, it is claimed that all great advances in physics are made by "physical intuition" rather than mathematical rigor and sophistication. This attitude, probably more than any other, is what distinguishes many theoretical from mathematical physicists. Let me briefly examine the two points in turn.

As for the first, it is certainly true in today's physics that almost all calculations leading to a direct comparison of theoretical predictions with experimental results are approximate and rarely employ rigorous mathematics. The great advantage theorists of the more phenomenological variety have is that they can forge ahead with verifiable theoretical results long before mathematical physicists can get to them, if they ever do. However, there is also the undeniable fact that in many instances what is called an "approximation" may not be a real approximation to a solution of the equations of the theory at all. It may simply be an intuitive scheme that the author has devised on the basis of the equations; the only reason it is thought to be a "good approximation" is that it agrees approximately with an experimental observation. There is no way of telling whether this "success" stems from the fact that the theory and the author's scheme are both right or both wrong in compensating ways, unless there is a sound mathematical demonstration that the method does, in fact, approximate a solution of the equations in some appropriate sense. In that case, we are back to asking a mathematical question.

Though rigorous mathematics may not be as useful for obtaining experimentally verifiable predictions from a given theory as the usual "approximation" schemes, it is much more useful for determining whether a theory or a solution method is mathematically viable and, if not, for suggesting a new method. The treatment of the Schrödinger equation for three particles is a case in point. Mathematical physicists demonstrated that the procedures used very successfully for two particles were not reliable for three or more, and a new set of equations was derived by the Russian Ludwig Faddeev. These Faddeev equations and their generalizations and extensions are now the methods of choice for few-body problems in quantum mechanics.

With regard to the second point, there is, no doubt, some truth to the contention that "physical intuition" rather than mathematical rigor is what leads to most of the important advances, prime examples being Rutherford, Fermi, and the early Einstein. In the introduction of the relativistic Lorentz transformation, Poincaré was a precursor of Einstein, but he was not enough of a physicist to realize and exploit the physical implications of that mathematical transformation. (Those who denigrate the value of mathematicians working in physics often use this example to show that they lack the physical intuition to make their work fully relevant.) This truth is by no means universal, however; neither Maxwell, who conceived elaborate models of little physical relevance that were later discarded, nor Dirac, who was guided largely by his sense of beauty, found their famous equations by physical intuition.

It is good to keep in mind that often what we call "physical intuition" is based on a particular mathematical technique. Of all the things Richard Feynman contributed to physics, the most memorable was the invention of the *Feynman diagrams* (see Figure 6). Most of our intuitive pictures in quantum field theory, especially in QED, are based on his graphs depicting the motion of particles—for example, electrons—and their emission and absorption of other particles—photons, for example—or the creation of pairs of particles and antiparticles. There is no book on modern quantum field theory, whether it is addressed to physicists or nonscientists, that does not expound the view of the vacuum as a cauldron of successive creations and annihilations of "virtual" particles, pair creations, etc., imagined

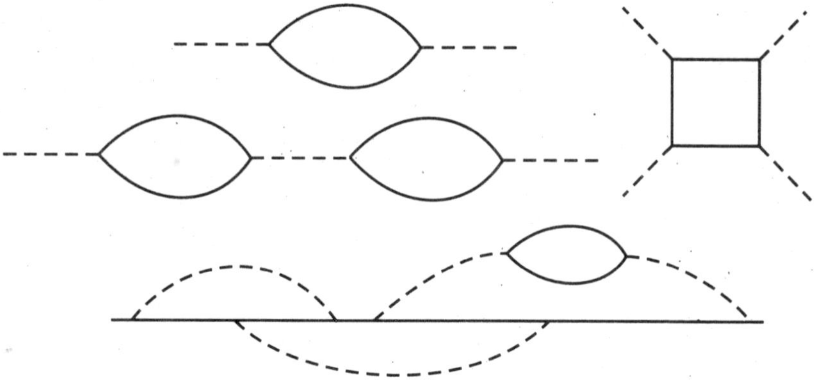

FIGURE 6 Feynman diagrams of the creation of electron-positron pairs by photons, and of virtual photons by electrons or positrons. The dotted lines represent photons, and the solid lines either electrons or positrons.

in terms of hosts of Feynman diagrams. All of these appealing images are nothing but pictorial representations of a particular method of solving the field equations, namely perturbation theory—successive approximations, taking smaller and smaller terms in the theory into account—even when the diagrams are intended to be modifications of that procedure. Had we known how to solve the nonlinear equations of the quantum theory of fields without using a series of perturbations of one kind or another, we would never have resorted to these pictures. (In fact, for "strong-coupling" theories, in which the successive terms do not become smaller and smaller—that is, all field theories except low-energy QED—they really make no sense at all.) Physicists use Feynman diagram heuristically and base much of their intuition on them all the same.

Poincaré presents another good reason for the pursuit of rigorous mathematics in physics. "On what condition is the use of hypothesis without danger?" he asks:

> The firm determination to submit to experiment is not enough; there are still dangerous hypotheses; first, and above all, those which are tacit and unconscious. Since we make them without knowing it, we are powerless to abandon them. Here again, then, is a service that mathematical physics can render us. By the preci-

> sion that is characteristic of it, it compels us to formulate all the hypotheses that we should make without it, but unconsciously.[4]

In other words, since it is one of the characteristics of rigorous modern mathematics to state explicitly *all* the assumptions under which a theorem is proved, the use of its strict methods minimizes the possibility that a tacit assumption sneaks into a theory or its experimental verification.

The Influence of the Computer

Since the Second World War a powerful new tool has been introduced into science and mathematics—the electronic computer. Because the use of such fast, large-scale calculating machines makes it possible to solve equations that were previously intractable, this new instrument has transformed areas of science as well as mathematics and opened them up for productive investigation. Even though the computer can rarely provide definitive proofs, it can often supply fruitful clues to suggest new effects and stimulate new ideas.

The study of nonlinear differential equations, touched upon earlier, is a case in point. The equations of classical mechanics are almost always nonlinear and therefore, in practice, unmanageable, with only some very special solutions either numerically calculable by hand or qualitatively comprehensible. Poincaré was the first to understand this well and to prove that most systems of classical particles evolve into chaos.[5] In the meantime, the fact that Maxwell's equations of electromagnetism and those of quantum mechanics are linear led mathematical physicists for nearly a century to pay almost exclusive attention to linear phenomena. For similar reasons, mathematicians concentrated primarily on linear equations rather than on the much larger body of nonlinear ones.

This situation changed when the introduction of the computer made the study of nonlinear dynamical equations, with their quasi-universal development of chaos, numerically feasible. Consequently, a fertile area of research emerged, directing the attention of mathematical physicists again to nonlinear phenomena, attention that led to real mathematical studies, not just numerical ones. The mathematicians A N. Kolmogorov, V. I. Arnold, and J. Moser, for example, showed that the borderline between the few solutions of the equations of a general nonlinear dynamical system that are "regular" and those that are "chaotic"[6] is surprisingly fuzzy: when parameters are changed from values for which the system is understood to be quasi-periodic and "regular," no matter how it may start, the system does not immediately become chaotic but retains its smoothly predictable character for a finite range of parameter values until it finally succumbs to chaos for all initial conditions. Their results became an important guide in numerical computations.

As a direct outcome of the computer's development, the question of the stability of the solar system, which Lagrange had answered with his newly invented linear perturbation theory, has been re-examined without the use of linear methods. By means of large computers, it is now feasible to check the predicted behavior of the planets and their moons over several million years, and it was found that the stability of the orientation of the Earth's axis of rotation is assured only because of the presence of the Moon[7]—an intriguing result that is based entirely on numerical computations rather than analysis.

The almost immediate influence of the computer upon the development of mathematical physics was exemplified by the investigation of the effects of nonlinear interactions between oscillators on a lattice, known as the Fermi-Pasta-Ulam problem. Solutions obtained for this problem by computers came as a great surprise and stimulated further interest in numerical computations for other nonlinear problems, such as the Korteweg-de Vries equation, which led to the discovery of solitons (mentioned earlier in Chapter 3). Similar results have been found for many other nonlinear equations, and their mathematical analysis, stimulated by numerical studies, has given rise to an extended branch of applied mathematics as well as to an explanation of a number of natural phenomena in plasma physics, biophysics, nonlinear optics, and many other areas. Thus the electronic computer sometimes serves a purpose in mathematics that is akin to that of an experimental device in physics—it leads to a new direction of thought by indicating unexpected mathematical effects that require an explanation. But this explanation cannot be furnished by the computer; it requires mathematical analysis. Even with the help of the most powerful computers available, mathematical physicists have so far been unable to overcome the resistance offered by a highly nonlinear phenomenon of great practical importance like turbulence.

The Relevance of the Nature of Mathematics

Mathematics is certainly a potent calculational tool and a convenient language for abstract concepts, but it is vastly more than that. Einstein went so far as to declare that "the creative principle [of science] resides in mathematics."[8] If mathematics cannot tell us any facts about the world, and it surely cannot, then why is it so invaluable for science? Poincaré's insight is a first step toward an answer: the greatest objective value of science lies in the discovery, not of things or facts, but of *relations* between them. "Sensations are intransmissible . . . But it is not the same with relations between these sensations . . . Science . . . is a system of relations . . . it is in the relations alone that objectivity must be sought."[9] Mathematics is the most appropriate vehicle for the *description of relations and for their logical exploitation.* Physicists do not use it (other than numbers) for the description of experimental facts, but for the manipulation of the interrelations between these facts—the theories. Mathematics "is only the means by which we use one set of facts to explain another," writes Steven Weinberg, "and the language in which we express our explanations."[10] The power of mathematics resides in its versatility in dealing with an enormous variety of connections between things, concepts, and ideas. I therefore cannot agree with the view of John Ziman that "physics defines itself as the *science devoted to discovering, developing, and refining those aspects of reality that are amenable to mathematical analysis.*"[11] Physics deals with *relations* between aspects of reality, and relations are always amenable to mathematical analysis. Whenever new kinds of relations do not seem to lend themselves to such manipulation, invariably a fresh branch of mathematics is discovered (or invented) that is suitable.

We are still left, however, with the question of why most branches of mathematics, often developed with no thought of application, have turned out to be so valuable in science, and especially in physics. "The enormous usefulness of mathematics in the natural sciences is something bordering on the mysterious and . . . there is no rational explanation for it,"[12] Eugene Wigner declared in awe. To try to solve that mystery we have to look, first of all, at the nature of mathematics.

Many practitioners regard proving a new theorem as *discovering* something that has an independent existence—mathematicians do not invent their structures but discover them. This school of thought, called *Platonism*, has a long history in the discipline, and many of the greatest mathematicians were, and are, its adherents. Some of them have found Platonism a great inspiration in their work—it is exhilarating to know that when you arrive at a long-sought insight, you have discovered a new piece of knowledge about the universe—a universe, of course, not of the external world of the senses but of Plato's realm of concepts and enduring truths. In this view, the work of mathematicians is quite analogous to that of experimental physicists, except that its scene of action is not Nature, as it is for scientists, but the eternal world of ideas. "I believe," wrote the English mathematician G. H. Hardy, "that mathematical reality lies outside us, that our function is to discover or observe it, and that the theorems which we prove, and which we describe grandiloquently as our 'creations' are simply our notes of our observations."[13]

The branch of mathematics in which this Platonic view is most convincing is the theory of numbers, which deals with nothing but the most basic entities of mathematics, the natural numbers—the integers. Since it is hard to believe that the integers are human inventions—they must surely be used even by an alien civilization on a planet of Alpha-Centauri—theorems about the prime numbers, for example, must have an independent existence, which mathematicians *discover.* "The integers were made by God: all else is the work of man," declared the German mathematician Leopold Kronecker, and he followed the precept of the second part of his assertion by arguments that led to the school of *intuitionism.*[14]

For the intuitionists, mathematicians are architects and engineers rather than explorers; their theorems are of their own making, and the tools they can use are correspondingly limited to those appropriate for construction. They must not make use of indirect inferences, such as proving a proposition by demonstrating that its denial would lead to a contradiction. During the early part of this century, the intuitionist school of mathematical thinking became fairly influential and had a number of prominent adherents. It also had passionate enemies, who opposed the implication that a number of important mathematical theorems, proved by methods whose validity the intuitionists denied, would have to be sacrificed. Intuitionism "seeks to break up and to disfigure mathematics," bitterly complained David Hilbert.[15] For example, the classical proof that there are infinitely many prime numbers is indirect: the assumption that their number is finite leads to a contradiction. Intuitionists deny the validity of this proof and will accept only a direct, constructive demonstration.

What is a Proof?

It seems clear now that the question "What is the nature of mathematics?" is not an idle one; its answer can have far-reaching consequences within and outside of mathematics and raises, at the same time, the related question "What constitutes a proof?" Mathematics, as we understand it, originated in the ancient Greek civilization; there, and only there, do we find the idea of an ironclad *proof* of a proposition. Many instances of the theorem of Pythagoras had been known to the Babylonians and the Egyptians, who made extensive use of it in practical land surveillance, for it allowed them to construct a right angle with pieces of string. But it never occurred to them to offer a general proof, a demonstration of its validity that would have to be accepted by every intelligent person who thought it through, a requirement that forms the basis of all of modern mathematics.

Would modern science be possible without a mathematics that relies on proofs? In his interesting book, *Pi in the Sky,* the astronomer John Barrow spins a tale of NASA at long last making contact with a highly developed alien civilization of great scientific and technical accomplishments. The Earth's mathematicians are waiting with bated breath for the powerful mathematical results such a civilization must have produced. Indeed, the aliens did know all the theorems in our mathematics books and more, but it turned out that what they called mathematics is not what we call by that name—they never developed the concept of a *proof,* but simply scanned large numbers of special cases using their very fast and powerful computers; if everything checked out, a "theorem" would be established. Since they looked at mathematics as a branch of science, in which a large number of correct predictions would be regarded as establishing a theory, so a large number of agreements with computer calculations could provide the needed evidence for the correctness of a theorem; if later an instance was found in which the theorem failed, it would simply be discarded like a superseded theory. Some of their philosophers occasionally wondered about the need for a procedure by which the correctness of a theorem could be established once and for all, but they were ignored because the mathematicians, realizing that such procedures would be very time-consuming, feared a slowdown of their progress. Such might well be the state of our mathematics if it had not been for the ancient Greeks.

The areas of *applied* and *pure* mathematics today are everywhere divided by a corrosive tension. Because applied mathematicians often do not follow the same stringent and rigorous demands that pure mathematicians make on their proofs, they are sometimes looked down upon by their *pure* colleagues, who claim that their inferior brethren do not *prove* their propositions. Since, however, applied mathematicians obtain results that are likely to be more directly useful in science, could they be the model for Barrow's aliens?

In answer, I would note first of all that Barrow's tale is highly implausible. The idea of theorems checked by computer calculations may sound reasonable in the discipline of number theory, where the computer would simply serve to test the validity of a proposition up to some very large integer. In most other parts of mathematics, however, and especially in analysis, which deals with the properties of functions that cannot be enumerated, this procedure would be inherently more problematic. But Barrow's tale is implausible chiefly because many branches of mathematics had their origin in ideas that arose in the course of long and arduous proofs of apparently unrelated theorems. Mathematicians put a high premium on fruitful proofs that contain *new ideas*, and they disdain demonstrations that are nothing but straightforward, if complicated, strings of logical steps. It is extremely unlikely, therefore, that a mathematics lacking proof requirements would have developed into the rich and varied intellectual area of thought that it is, and I think most applied mathematicians, even when feeling harassed by their more rigorous colleagues, would agree.[16] Theorems that were proved by less stringent methods were sometimes found later to be false in their original form, which usually meant they were valid only under hypotheses more restrictive than had previously been thought. It is precisely the discovery of such exceptional cases, whose existence had not been anticipated, that often led to new concepts and opened up new fields.

Poincaré gives us an additonal reason, quoted earlier, for believing that a mathematics without real proofs would provide a poor structure for science: namely, that the exacting requirement to list all the hypotheses needed for the proof of a theorem is an extremely useful safeguard for science, one that prevents unconscious assumptions from entering surreptitiously into its theories. For all these reasons, I am of the firm opinion that modern science would not be in the highly deleveloped state that it is today without the kind of mathematics that requires general proofs for its propositions rather than merely verifications in specific instances.

Whence Comes the Power of Mathematics?

We are, however, still far from the point to be explained—why are the results obtained by mathematicians, who often work without thought to any application of their ideas to the real world, nevertheless almost invariably found to be useful? One answer is given by those in the field who regard what they do as inventions of the human mind and, thus, the result of biological evolution. Such was the point of view taken by the founder of the school of intuitionism mentioned earlier, the Dutch mathematician Luitzen Brouwer. For the followers of this school, there is no real problem here. If mathematics is a product of our brain as it developed in evolution, then that brain—and also mathematics—can be expected to be automatically adapted to the needs of its environment, that is, Nature. This argument does not convince me, because the parts of Nature that might be imagined to have had an influence on the development of the human mind are but a small portion of what the theories that are couched in the most abstract, mathematical language are designed to explain. In order to believe that the mathematics needed for the formulation of string theory or of the general theory of relativity is a natural outcome of the evolutionary development of the human mind, we would need evidence that the brain is influenced by particle physics at enormously high energies or by much stronger gravitational fields than encountered on Earth. The leap from the nature of evolutionary pressures to the flights of mathematical imagination is too great for such a notion to be plausible.

In my view, Freeman Dyson has come closest to an understanding of the mystery of the "unreasonable effectiveness of mathematics." After quoting the Austrian physicist Ernst Mach, who said with great acuity that "the power of mathematics rests on its evasion of all unnecessary thought and on its wonderful saving of mental operations," he observes that "a physicist builds theories with mathematical materials, because the *mathematics enables him to imagine more than he can clearly think.*" He goes on to explain that

> the physicist's art is to choose his materials and build with them an image of nature, knowing only vaguely and intuitively rather than rationally whether or not the materials are appropriate to the purpose. After the design of the theory is complete, rational criticism and experimental test will show if it is scientifically sound. In the process of theory building, mathematical intuition is indispensable because the "evasion of unnecessary thought" gives freedom to the imagination.[17]

But Dyson also immediately issues a warning that reverberates among those scientists who are suspicious of too much mathematics—"mathematical intuition is dangerous, because many situations in science demand for their understanding not the evasion of thought but thought."

I believe that the reason mathematical results and ideas are as useful and powerful in science as they are is, in the end, not so very mysterious. The unraveling of the enigma of Nature demands all the intellectual power humans can muster, and possibly more. They must therefore avail themselves of all the logical tools at hand, and by far the most powerful of these tools is mathematics. It is not that Nature's own language is mathematics—as Galileo thought—and that we are thus compelled to learn every obscure rule and usage of that tongue to comprehend it, but that mathematics is *our* most efficient and incisive instrument for rational understanding of relations between things. If mathematicians have already built, with great ingenuity, elaborate structures containing results of long and hard thought, if they have devised concepts appropriate for reaching their conclusions, then scientists are only too happy to make use of this "wonderful saving of mental operations." If they have not, scientists will build these structures themselves, or at least build their skeletons and leave the fleshing out to mathematicians. Dirac invented or discovered the mathematics he needed on his own when necessary; his so-called delta function is an example of a concept, very useful and productive for physics, that was at first derided by mathematicians and later made mathematically kosher. In sum, *mathematics serves as an immensely efficient set of mental shortcuts.*

The theory of groups, with which this chapter began, may serve as an illustration of the economy of thought achieved by mathematics. Many important predictions of contemporary physical theories are consequences of the existence of symmetries in these theories—symmetries manifesting themselves in ordinary space and time or in more abstract spaces. Their mathematical formulation is embodied and their consequences codified in the concepts and theorems of group theory, the branch of algebra initiated by Evariste Galois. Conceivably, the resulting predictions could be made without using the concepts and results of the theory of groups; at the very least, however, to arrive at them in such a way would take enormously more time and effort. It is not that Nature itself makes use of group theory; it is that we humans need this invaluable mental crutch to understand Nature. Mathematics is not embedded in the structure of reality, but we require the help of its power to penetrate and describe that reality.

Nevertheless, I have to admit that my arguments do not really answer the puzzling question why almost all mathematical structures, even those erected purely for the sake of their admirable architecture, eventually turn out to be useful tools for physics. One might almost be tempted to ascribe it, as Einstein did, to a Leibnizian "pre-established harmony" between our thought processes and Nature.

8

CAUSALITY, DETERMINISM, AND PROBABILITY

JUST as a child, wanting to understand what makes her toy work, wiggles one piece to see how another piece waggles, so a physicist in his laboratory, looking for an understanding of Nature, effectively conducts experiments by jiggling one part of Nature and watching how other parts respond. In science, we are always searching for cause-and-effect relations. If the principal motivation of scientists is to find explanations, causes have been the primary explanatory principles from well before Aristotle.

Efficient Causes

Aristotle's philosophy offered four kinds of causes, *efficient, final, material,* and *formal,* of which the last two are now archaic. We no longer regard steel and glass as a (material) cause of a telescope, and we do not think of the conservation of momentum as the (formal) cause of the damage in a car crash. Until the nineteenth century, the aim of biological taxonomy was to discover the ideal Platonic structures underlying the observed diversity of life, thereby revealing their formal cause. The abandonment of Aristotle's final cause, directed toward a purpose, is more recent, but its hold on people's imagination remains strong. To the extent that there was a theory of evolution before the nineteenth century, it endeavored to explain natural history by Nature's striving for perfection, a final cause; Darwin's revolutionary theory replaced both the formal and the final cause by an efficient one: natural selection. Only efficient causes, and their historical successors, are of real interest to us today.

The power of efficient causes lies in their ability to *compel* their effects to come about; it is as though they act with an anthropomorphic sense of effort—or through love and hate, as Empedocles of Acragas, who originated this causal notion, saw it. (He was the legendary Pythagorean who, wishing to demonstrate his immortality, is said to have ended his life by jumping into the crater of Mount Etna.) There can be little question that this idea influenced Newton's formulation of the laws of motion, which relied heavily upon the concept of a force as the efficient cause of acceleration. It also lay at the core of the general philosophical resistance encountered by his concept of gravity, a force described as acting at a distance, with no agent to transmit the causal compulsion.

The Aristotelian notion of efficient causes, as it relates to science, was effectively overturned by David Hume, who argued convincingly that no observation of the relation between cause and effect could possibly lead to the conclusion that a compelling power exists or that there is a *necessary* connection between the cause and the effect; all that can be observed was a constant conjunction between the two. On purely empirical grounds, to say that *A* causes *B* means no more than that whenever *A* happens, *B* occurs. To Kant, however, this radical proposition appeared to be corrosive of science, in which, he thought, it was crucial for a cause-effect sequence to be governed by a universal rule. His epistemology, designed to rescue science from Hume's skepticism, made causality a category of rational thought—not a necessary property of Nature, but an indispensable way of understanding it. "Denying causality," writes Popper a century and a half after Kant, "would be the same as attempting to persuade the theorist to give up his search."[1] The essential insight that science does require the concept of causality in some form has survived Hume's destruction of Aristotelian efficient causes, and in this chapter I want to trace the use of the notion of causality in science through its vicissitudes over the last two hundred years, ending with its turbulent fate in the quantum theory.

Although, when pressed, most physicists today would no doubt side with Hume and disavow any notion of efficient causes in the Aristotelian sense, it is hard to deny that when they look for an explanation they are searching for something more than a constant conjunction or, in a weakened modern form, a statistical correlation. The many contemporary controversies concerning ecological or medical phenomena—think of the connection between smoking and lung cancer—show how little power to convince mere correlations wield. The great advantage physics has over astronomy and large parts of biology is that experimentation establishes causation much more persuasively than mere observation does. Ultimately, we are usually not satisfied until we have found, in one sense or another, a mechanism that is responsible for the causal connection. Einstein's explanation of what we mean when we say that we understand a group of natural phenomena—"that we have found a constructive theory which embraces them"—shows that understanding for him meant finding a cause which included a mechanism; the "theories of principle," which might be said to furnish a formal cause in Aristotle's sense, did not lead to real understanding.

How do we establish by experimentation that *A* causes *B*? Schematically, we turn on *A* and see if *B* happens. But if we switch *A* on at regular intervals and B occurs equally regularly, perhaps *B* happens by itself at just those times. In order to exclude such accidental coincidences, we must vary the occurrence of *A* at will (or perhaps according to a random device like a roulette wheel); only then can we be sure that the correlation between *A* and *B* is not fortuitous. The essential point is that it is important for the times (and the nature) of *A* to be *under our control.*

The dividing line between passive observation and active experimentation can sometimes be blurred. In high-energy physics experiments, for instance, the primary particles and their energies are chosen by the experimenter, who also designs the nature and arrangement of the apparatus that detects the products of collisions between the primaries. That, however, ends the experimental control. Since some of the emerging particles are produced immediately and others in secondary collisions or decays, there is a lack of detailed control over the intermediate steps leading to the final product. To deal with such situations, experimenters often resort to computer simulations by means of "Monte Carlo" programs, which make models using random choices of data to see how the detectors would respond if the purported cause were absent.[2]

Suppose we live in a totally deterministic universe and all our perception of control is illusory? In such a universe a cause-effect relation can never be definitively established. If *everything* runs like clockwork forever, saying that *A* causes *B* has no meaning, since this presupposes the possibility that *A* might not happen. There is no way to tell if the clockwork is externally directed or self-propelled—we could never find out if we are puppets or automatons. Such a world can produce only history, for science can function only if we act *as if* we had at least some control over causes.

Time Delay

An important property of the relation between causes and effects is the delay between them; though this time interval may be vanishingly small, we all know that effects never precede their causes. The question is, do effects follow their causes *by definition*, or is this a fact of Nature that we have learned from experience? Some parapsychologists claim the existence of precognition, purportedly established by experiments in which an effect preceded its cause. Their demonstration would be set up just as previously described: *A* is governed by our free will or by a random device—Susan throws a pair of dice—and its occurrence is perfectly correlated with that of *B*—Peter writes down the results of the throws—only in this case *B* happens before *A*. Faulty and unconvincing though these experiments are, since there is nothing *logically* inconsistent in them we have to conclude that effects never precede their causes as *a matter of experience*. The temporal order of cause and effect is a fact of Nature, not a logical necessity.

Consider what the consequences would be if there were, in fact, some kinds of effects that preceded their causes—that is, if precognition were possible. In that case, we could send a signal into our own past: event *A* at the time t_1 causes an event *B* at the later time t_2, and the occurrence of that event causes event C at the time t_3 before t_1, in turn causally influencing *A*. This is sometimes called a *causal cycle*. Imagine now that event C is such as to prevent *A* from happening[3]—it detonates a bomb, destroying the building in which *A* occurred. As a result, *A* could not have happened, and hence neither would *B* and *C*, which in turn would make *A* possible again. Have we arrived at a *logical* contradiction, which would make precognition self-contradictory? No, because it is conceivable that every time we try to start the cycle by switching on *A*, it simply will not work—the switch won't budge. But such a consistent malfunctioning of an otherwise working mechanism is quite contrary to all our experience, indicating again that the causal cycle confronts us not with a logical contradiction but rather with a denial of a large body of experience.

I want to emphasize that to say the temporal order of the cause-effect relation is a matter of experience is not at all to imply that there is any reason to doubt its universality. Because we have such an enormous amount of empirical evidence demonstrating that effects never precede their causes, we may regard this as an extremely well-confirmed matter of fact and use it to define an arrow of time in Nature.[4]

What of time, then? Is it an illusion, and could physics just as well do without it? Even though there are scientists who do envision a timeless—not static!—cosmos, whose entire history would be laid out before us at once, the concept of time is a crucial element of all physical science. The very idea of scientific explanation and knowledge on the basis of experimentation is founded on causality and the time-order of cause and effect. It is not an accident that scientific laws are mathematically expressed in the form of differential equations with respect to time: they incorporate the notion that a general theory explains what happens *now* in terms of occurrences either at the preceding or at the subsequent instant. In accordance with the causal time-order, we usually solve such an equation on the basis of the *initial* situation, thereby converting the law into a hypothetical description of history, flowing from past to future.

Use of Causality in Classical Physics

In the world of concrete science, as opposed to more philosophical endeavors, the use of the arrow of time has significant consequences. There are specific instances in physics in which causality-related arguments play an important role in selecting which of all the possible solutions of the equations of a theory is the right one. Let me describe a prominent example.

The behavior of moving electrically charged particles in interaction with one another, together with the electromagnetic fields they produce and react to, is classically described by the Maxwell-Lorentz equations. These are partial differential equations, and their solution requires the specification of boundary conditions and initial conditions, stating how the fields behave on the boundary of the region under consideration and how they began. Consider, then, the mathematical problem of determining the electromagnetic field of an electrically charged point particle, made to move along a prescribed trajectory at a given variable speed. There are infinitely many solutions of the Maxwell equations for this situation, but it is traditional to use an argument of causality to select the one of physical interest. If we assume that before the particle began to move, the only field present was the static electric field of the point charge at rest, which falls off like the inverse square of the distance, the field is uniquely determined after the motion starts. The solution thus singled out is called *retarded,* because the field at any given point P at the time t is determined by the acceleration of the charged particle at an earlier time, retarded with respect to t by the length of time it takes for light to travel in a straight line from the position of the charge to the point P. Since it makes good causal sense, this "retarded field" is regarded as the only solution physically acceptable, and it leads to the conclusion that a moving electric charge *emits* electromagnetic radiation, which travels away from the charge. Without the causality argument, the solution would not be uniquely determined. Indeed, there is also an *advanced* solution, which features radiation traveling toward the charge; radiation in this case, present even before the particle began to move, is of a strength depending on its acceleration at a *later* time.

As in the case of the second law of thermodynamics in statistical mechanics, the causality argument, here used for the definition of radiation, introduces an element of *irreversibility.* The gas molecules emerging from an open bottle into vacuum could not be returned to it because their positions and momenta would have to be controlled with a precision impossible to achieve in practice. Likewise, the radiation that arrives on a closed surface surrounding, at a large distance, the emitting electric charge is practically impossible to send back converging to a point because the precise values of the waves at various distant positions would have to be mutually correlated to a degree that is impossible to produce precisely. The emission of radiation, like the escape of a gas from an unstopped bottle, is thus an irreversible process.

Some fifty years ago, Richard Feynman and John Wheeler put forth an ingenious theory utilizing a symmetric combination of the retarded and advanced solutions instead of just the retarded solution of the Maxwell equations. This theory postulates the existence of a giant absorber at the edge of the universe, which soaks up half of the radiation reaching it, with the same net result as that of the conventional theory. Its primary virtue is that, at the expense of violating intuition, it avoids the usual causality argument, but it has not found much acceptance among physicists.

Another application of the causality concept, with important consequences, occurs in the theory of relativity. Einstein derived the *Lorentz transformation,* which relates the spatial coordinates and the clock times in one reference frame to those in another frame moving at a constant velocity with respect to the first, on the basis of two assumptions: (1) the principle of relativity—the laws of physics have the same form in all inertial frames of reference,[5] with no laboratory singled out as being absolutely at rest; and (2) the constancy of the speed of light—the velocity of light is the same in all such laboratories.

One of the consequences of this transformation is that if a signal moving faster than light were sent from point *A* to point *B*, there would always exist another frame of reference in which the same signal would start at *B* and end at *A*—the time of the signal's coincidence with *B* would be earlier than its coincidence with *A*. Therefore, if such *superluminal* signals were possible, the temporal order of a cause-effect relation would depend upon the observer's frame of reference. If we accept as a universal empirical law that causes always precede their effects, this would mean that what appears as the cause to one observer would be regarded as the effect by another, and *vice versa,* contradicting the definition of the cause as being under our control. Imagine an order to commit murder being sent from Smith to Jones, and another observer seeing the order going from Jones to Smith. Who is the criminal? Superluminal signals could also be used to send messages into your own past, and causal cycles of the kind described earlier could be set up,[6] with all their unacceptable consequences. Einstein therefore concluded that superluminal signals must be physically impossible. In other words, on the basis of the theory of relativity, together with our accepted notion of causality, he was able to infer a very general stricture on Nature: there can be no signals of any kind that move more rapidly than light—no energy can be transported, no information sent, no influences transmitted, faster than the speed of light. As a result of this prohibition, the word *causality* in relativistic physics has come to mean exactly that: not only do effects follow their causes, but if two causally connected events occur at different locations, there must be a long enough time lag between them for a light signal to reach one from the other.

Relativistic causality in this sense has quite concrete consequences in quantum field theory. In quantum mechanics, physical observables are described by *operators,* mathematical objects that not necessarily follow the rules of numerical algebra according to which $a \times b = b \times a$. If two observables represented by a and b have the property that $a \times b$ differs from $b \times a$, one says that *they do not commute,* and this lack of commutativity is at the root of Heisenberg's indeterminacy relation—the observables represented by a and b cannot both be measured simultaneously with unlimited precision. Since this indeterminacy can be interpreted as an influence on one of them by the measurement of the other, and relativistic causality requires that such an influence cannot travel faster than light, relativistic quantum field theory contains a fundamental requirement that all pairs of variables representing observables (including the mathematical operators representing the fields) referring to two different times, t_1 and t_2, and to two locations farther apart than light can reach during the time interval $t_2 - t_1$ must necessarily commute. In this manner causality puts a severe constraint on the structure of any relativistic quantum field theory that may be proposed as a potential description of Nature; theories violating it, called "nonlocal," are generally shunned.

Determinism and the Definition of State

The doctrine of determinism holds that everything in the world is completely determined by prior causes. This is what Laplace meant when he said (as quoted in Chapter 7) that, to an "intelligence" that could comprehend all the forces of Nature, "nothing would be uncertain and the future, as the past, would be present to its eyes." What was the basis of this striking image of the universe as predictable as clockwork?

In the nineteenth century, Newton's equations of motion were reformulated, without changing their physical content, by the Irish mathematician William Rowan Hamilton in a way that simplifies their analysis. Hamilton's equations for a system of particles deal not just with the positions of the particles but also with their *momenta*. In contrast to Newton's equations for the positions of the particles, Hamilton's equations for the positions and momenta are differential equations of the first order and can be uniquely solved by specifying the initial values of all the locations and velocities. Therefore, if we know the positions and momenta of all the particles (and all the forces between them) in a closed system—for example, the universe—at one time, they are determined for all future times. This is the essential content of Laplace's dictum.

Another way of formulating the determinism embodied in Newton's laws of motion is to say that if we know the *state* of the universe at one time, the laws allow us to predict its state at any future (or past) time. This version is equivalent to Laplace's pronouncement if the word *state* refers to the combination of the positions *and* the momenta of all the particles. (In the latter part of the nineteenth century, Maxwell added the electromagnetic fields to the description, which meant that the state of the universe had to include the electric and magnetic fields as well, but the deterministic statement remained correct.) Therefore *what we mean by the state of a physical system depends on the structure of the dynamical equations describing the motion of the system.*

To determine the state of a system at a given time, we have to choose a maximal number of independent physical parameters—which depends on the equations of motion—that specifies the system completely. If the equations of the theory are such that these parameters, given at one time, determine their values at all later times, the theory is deterministic.

In Newtonian mechanics, if the state of a group of particles were defined by their positions only, there would be no determinism—the future state would not be determined by the earlier one. If, alternatively, we included in the state their accelerations, the initial state could not be freely specified, because the equations of motion do not allow us to choose the positions and accelerations independently—given the forces and the initial positions of the particles, Newton's equations directly determine their initial accelerations. On the other hand, imagine living in a universe governed by Aristotelian mechanics, in which the forces determine the *velocities* of all particles—if we lived immersed in a highly viscous medium, such would, in fact, be our world. In that case, Laplacean determinism would hold sway as well, but the state of the universe would be specified by the positions of all the particles, rather than by the positions and momenta together, because the Aristotelian equations of motion would make the momenta depend directly upon the positions.

The notion of the *state* of a physical system, therefore, should not be regarded as either intuitively obvious or determined *a priori*. Classical physics is deterministic, with a very specific definition of the concept of state dictated by its laws of motion.

The Quantum World

The most famous, or notorious, property of the quantum theory is its purported abandonment of determinism. Whereas the classical laws allowed us to predict the motion of particles with unlimited precision, the quantum theory can give us only probabilities. A very exact and well-verified law predicts the probability that a radioactive atomic nucleus will decay in any given length of time—but we cannot predict when an inividual nucleus will do so. Listen to the clicks of a Geiger counter near a piece of radioactive material, and you will hear the evidence for their randomness. It was the appearance of acausality in decay events that induced Einstein to utter his famous complaint that he could not believe God played dice with the world. The universe, it seems, does not follow Laplace's dictum, after all—a blessing to some and a misfortune to others. In fact, however, the situation is much more complicated.

The quantum-mechanical state of a system of particles, if specified as precisely as it can be, is described by its *state vector* or *wave function*.[7] The place of the choice of a maximal number of independent parameters, as is done to specify a classical state, is here taken by what is known as the *Hilbert space* to which the wave function belongs. The development of this wave function in the course of time is subject to the Schrödinger equation, a differential equation of the first order like Hamilton's equations in classical mechanics. This equation uniquely fixes the value of the wave function at any future time if it is given at the initial instant. Therefore, just as in classical physics, if the state of a system is specified at the beginning, its future state at any later time is fully determined—but the specification of the quantum-mechanical state does not contain as much information as does that of a classical state. The initial state, for example, may be one in which all the particles have precisely given positions, or it may be one in which they have precisely given momenta, but it cannot be one in which both the positions and momenta are precisely given. The state at a later time, though completely determined, does not, in general, allow us to compute the precise locations or momenta—it gives us only probabilities for finding the particles in any given positions or with any given momenta.

Some have found it puzzling that the quantum theory forbids certain specifics in its definition of the state of a system—the quantum theory "does not allow certain questions to be asked"—whereas it is thought that classical physics does not forbid anything. This reasoning, however, reveals a misunderstanding. As we have seen, in Newtonian mechanics we are also not permitted to specify at will all the properties of a system of particles we might want: we cannot arbitrarily set their positions and their accelerations. The boundaries of what physical attributes can be freely assigned are simply different in the quantum theory: in specifying the state of a particle we are not allowed to give it, as we could in classical mechanics, both an arbitrary position and an arbitrary momentum; but if we lived in an Aristotelian world, we could not do so either. In such a world, however, a particle's position would determine its momentum precisely, whereas in quantum mechanics it gives it a probability distribution.

The essential difference between classical and quantum physics does not lie in how deterministic the development of the physical state is but in the meaning of the word *state*. When the state of a system is given in classical physics, *all* dynamical variables—such as positions and momenta—have precise values; when it is given in the quantum theory, the values of *some* are assigned probabilities and, as a result, predictions of the development of these variables become acausal. The interpretation of the quantum theory, central to modern physics, therefore hinges on the meaning of *probability*, which we should now examine in more detail.

Probabilities

The ideas of probability and statistics first entered science in the middle of the nineteenth century, through the discipline of statistical thermodynamics, which is designed to describe the behavior of large conglomerates of atoms or molecules. When we study these large systems, such as gases and liquids, we are not concerned with the motion of each individual molecule; we are satisfied with a variety of averaging techniques to explain the effects of the motions of all the molecules on the overall properties of the fluid. In order to account for the statistics, Josiah Willard Gibbs introduced the notion of an *ensemble*, a large (in principle, infinite) collection of identical copies of the system, all moving independently. Probabilities that the system as a whole would take particular states were then computed as fractions of this ensemble.

The logical and mathematical concept of probability, on the other hand, is much older than its scientific adoption. Intuitively we all have an idea of what *probability* or *likelihood* means, but, except for the simple cases of a finite number of outcomes each of which has an "equal chance" of occurring, this idea is very difficult to pin down with any precision. For Aristotle, probability meant a *propensity:* to say that the probability of a coin coming up heads in a toss is 1/2 means that the coin has an innate tendency to land heads-up with a fifty-fifty chance. A quite different view was advanced early in this century by the Austrian mathematician Richard von Mises, for whom the probability of throwing heads was defined by the fraction coming up that way in an infinite sequence of tosses with the same coin, or in an infinite set of simultaneous tosses with identical coins; this is called the *frequency interpretation.* Adherents of the propensity theory would, of course, agree that the frequency of an outcome in many repetitions is a way of determining its probability, but they would deny that this is its *definition.* The fundamental difference between the two approaches lies in the fact that the Aristotelian propensity is a property of an individual system—a specific coin—whereas the frequency-defined probability is a property of a large (in principle, infinite[8]) sequence of identical systems.

The contingent nature of probability shows why the Aristotelian propensity theory is unsatisfactory. What is the meaning of the statement "The probability that Smith will die in 1998 is x percent"? Smith's life-insurance premium will depend on the value of x, and her insurance company must be able to calculate it. By itself, however, the statement is meaningless. The insurance company can com-

people. If the rate setters consider her simply as an American, the actuarial tables will show one value of x; if they know she is forty years old and lives in Chicago, the tables will give another value; if they also know she has cancer, a third value. In other words, the meaning of "The probability that Smith will die in 1998 is x percent" depends upon the (large) population of which Smith is considered to be a member; it is not intrinsic to Smith. This population determines the value of x in the actuarial tables, and this population makes up the set used in the frequency theory to give meaning to the probability. We might think that it is possible eventually to arrive at Smith's *intrinsic* propensity to die in 1998 by defining the set of which Smith is a member more and more narrowly, until eventually it can contain only duplicates of her with identical histories—but in a deterministic universe, that intrinsic probability could then only be 0 or 100 percent.

The man who believes that he can decrease the chances that there will be a terrorist with a bomb on board his plane by bringing a bomb himself, because the chances of having two bombs on the same plane are much lower than the chances of having just one, is making the error of thinking of probability in Aristotelian terms. The probability of a terrorist bomb on board is the ratio a/b, in an imagined large number of repetitions of flights of the same kind, of the number a of flights with terrorist bombs to the total number b of flights—if a includes only flights in which another, innocent passenger also brings a bomb, then so must b, and the ratio will be the same as when neither a nor b includes such flights.[9]

In most areas of science, the difference between the Aristotelian and the frequency notion of probability is of little consequence and can be ignored. When a physician tells a patient that his chances of recovery are 90 percent, the patient does not worry whether she means that the probability is attached to him as an individual or to a large group of patients. In fact, however, what she does mean is that extensive experience has shown that 90 percent of all people with the same medical condition do recover. When the weather forecaster announces a 50 percent chance of rain in your town for tomorrow, many listeners may interpret this to mean that there is a propensity of 1/2 for rain; in fact, it means that meteorologists expect rain to cover one-half of the area reached by the broadcast, though they don't know where. Therefore your town has a fifty-fifty chance of being in the area where it rains, and in many repetitions of such broadcasts, your town should get rain half the time. The exact meaning of probability statements is, for most of everyday life, of little concern; not so for quantum mechanics.

Consider the interpretation of the quantum-mechanical wave function describing the state of a system. This function embodies all the probabilities the theory predicts—given the wave function of an electron, we can calculate the probability of its being located in any given region of space as well as the probability of any other of its variable properties. Some physicists take the view that probabilities rather than certainty enter the theory because the wave function describes our *knowledge* of the system, which can never be complete. This interpretation, in which the consciousness of an observer plays an important part, introduces a strong element of subjectivity into physics, which many, including me, emphatically reject. Since the quantum theory offers no compelling reason for the adoption of philosophical idealism for those without an *a priori* favorable bias toward it, I shall confine myself to *objective* interpretations.

If probability in the quantum theory is taken as a propensity, it is attached to the individual electron whose changing behavior in the course of time is described by the wave function. As in classical physics, the state of the system "electron" is then a property of one particle. On the other hand, if the probability is considered as a frequency, the wave function represents an *ensemble* of particles—by this interpretation, quantum mechanics never deals with single systems but always with ensembles. This view therefore makes the quantum theory much more analogous to statistical mechanics than to Newtonian physics, which deals with identifiable individual systems.

These two slants on the quantum theory, and their connotations, are separated by a large gulf. In the first approach, calculating the wave function of an electron means finding its individual behavior; in the second, it means calculating the behavior of an ensemble of which the electron is a member. The frequency interpretation naturally—though not inevitably—leads to the question: what is the quantum analogue of the physical substratum that, in statistical mechanics, allows us to determine probabilities from its dynamics? In classical statistical mechanics, gases are assumed to consist of a large number of molecules, whose motions, governed by deterministic Newtonian mechanics, we ignore in detail but on the basis of which we calculate the probabilities and statistical properties of interest. As a result, we arrive at probabilistic, statistical statements despite the fact that the substratum—the collection of molecules—is governed by completely causal laws. In quantum mechanics, we do not postulate such a substratum but arrive at probabilities directly, making the theory acausal. Given the close analogy with statistical mechanics suggested by the ensemble interpretation of the wave function, however, might there not be a substratum below quantum mechanics, after all, to which we normally pay no attention but whose perfectly deterministic behavior leads to the probabilities at the quantum level?

This question has occupied a small number of physicists ever since the 1920s, and Einstein strongly supported their efforts to answer it. If the "hidden variables" of a hitherto unknown sub-quantum level could be found, the acausal nature of the quantum theory would become much more palatable, because it would simply indicate our lack of knowledge of and access to the hidden details. That is the case with the apparently erratic and acausal Brownian motion of dust specks seen under a microscope; we know the specks are buffeted by the much smaller, invisible surrounding molecules moving deterministically according to Newton's laws, but we cannot track these molecules. The "hidden variables" approach has not succeeded, however, and as we shall see in Chapter 9 there are good reasons for believing it cannot succeed unless the hidden substratum has properties that are just as counterintuitive and weird as the quantum theory.

Popper's Propensities

Where does this leave the use of probabilities in the quantum theory? An attempt to solve the dilemma of having either to use unacceptable, absolute Aristotelian propensities or to forgo describing the behavior of individual systems by means of quantum mechanics was made by Karl Popper. Seeking to attach probabilities to individual systems, he proposed a redefinition of propensities by a modification of the frequency theory. Since his theory has a large number of conscious or unconscious adherents among physicists, it is worth discussing in some detail.

According to Popper, the frequency interpretation "attributes to the single event a probability *merely* in so far as this single event is an element of a sequence of events with a relative frequency. As opposed to this," his non-Aristotelian

> propensity interpretation attaches a probability to a single event as a representative of a *virtual or conceivable sequence* of events, rather than as an element of an actual sequence. It attaches to the event a a probability $p(a,b)$ by considering the *conditions which would define this virtual sequence:* these are the conditions b, the conditions that produce the hidden propensity, and that give the single case a certain numerical probability.[10]

In contrast to Aristotle, Popper therefore gives propensities a contingent quality, but he wants to avoid the drawback in the frequency definition of probabilities being divorced from individual events or systems. "It will be asked why I propose to introduce hidden propensities behind the frequencies? My reply is that I conjecture that these propensities are physically real in the sense in which, say, attractive or repulsive forces may be physically real."[11] And he explains that "just as a field of force may be physically present even when there is no (test) body on which it can act, so a propensity may exist for a coin to fall heads even though it falls only once, and on that occasion shows tails. There may indeed be a propensity without any fall at all."[12] In Popper's view, the important analogy between the notions of force and propensity lies "in the fact that both ideas draw attention to *unobservable dispositional properties of the physical world,* and thus help in the interpretation of physical theory,"[13] but he draws a sharp distinction between his concepts and Aristotle's:

> Like all dispositional properties, propensities exhibit a certain similarity to Aristotelian potentialities. But there is an important difference: they cannot, as Aristotelians might be inclined to think, be inherent in the individual *things.* They are not properties inherent in the die, or in the penny, but in something a little more abstract, even though physically real: they are relational properties of the total objective situation; . . . In this respect, propensities again resemble forces, or fields of forces: a Newtonian force is not a prop-

> erty of a thing but a relational property of at least two things; and the actual resulting forces in a physical system are always a property of the whole physical system. Force, like propensity, is a relational concept.[14]

In the last analysis, Popper's propensities are, of course, closely related to frequencies. Indeed, propensities "are dispositions to produce frequencies . . . But propensity does not *mean* 'frequency,' for there are events too rarely repeated to produce anything like a a good segment of a random sequence (or a 'frequency'); yet these rare events may well have a propensity."[15] As you can see, there really is little difference between frequency-defined probability and Popper's propensity. After all, the frequency theory need not be interpreted to imply that an actual repetition or duplication of a system or event, in infinitely many copies, must be possible for it to be applicable. Such a literal interpretation has made cosmologists especially uncomfortable in dealing with the wave function of the universe—how could we even conceive of an ensemble of universes?—and has led some of them to avoid doing so by reformulating the quantum theory.[16] Popper's propensity concept shows that a reformulation is quite unnecessary, but if, finally, his notion differs at most *psychologically* from that of frequencies, it does have the advantage of enabling us to speak of the wave function of an individual system so that we can avoid being led to the analogy of statistical mechanics. Another way to escape this analogy would be to adopt an attitude of agnosticism toward the ultimate explanation of the probabilities. Perhaps this is what Popper had in mind when he mused that "propensities are as 'transcendent' or 'metaphysical' as Newtonian forces (which Berkeley denounced as 'occult')."[17]

To recapitulate: with respect to the crucial scientific concept of causality, the difference between classical physics and quantum physics cannot be drawn by saying that in the former the state of a system of particles at one time determines the state of that system at any future time and in the latter it fails to do so. In fact, the state of a system of particles at one time determines the state of that system at any future time in both cases. The striking difference is that in classical physics we can, in principle, tell with certainty every detailed property of a particular system of particles if we know its state, whereas in quantum physics, "every detailed property" is not specified in the same sense when we know the state: the parameters fully determined classically are quantum-mechanically determined only probabilistically. This means that if we insist on describing a real physical system in terms of classical variables—in terms of the positions and momenta of particles—the resulting description cannot be deterministic. Moreover, if the frequency theory of probability is taken literally, the quantum theory never deals, as the classical theory does, with one particular system; instead, a quantum *state* always describes ensembles. On the other hand, if we were to abandon classical pictures and choose an intrinsically quantum-mechanical formulation, the sub-micro world may well be deterministic and probabilities might never be needed. But this raises the question of what reality is and how it can properly be described, which I shall take up in Chapters 9 and 10.

9

REALITY ON TWO SCALES

IT is difficult to imagine a scientist who doubts that a real world exists independently of ourselves. We measure its properties, we observe its changes, we try to understand it, and sometimes it astonishes us. "The belief in an external world, independent of the perceiving subject, lies at the basis of all natural science," Einstein insisted. "Since, however, the sense perceptions give only indirect indications of this external world, that is, of the physically real, it can be conceived by us only by way of speculation. As a consequence," he concluded, "our concepts of the physically real can never be final."[1] Others draw more dramatic conclusions—what we know of the world out there rests on nothing but individual sense impressions; we hear the rustling of a tree's leaves, we see its branches, and we can climb it. Might the tree not simply *be* the sum total of these impressions, tied together into one metaphysical bundle? That is, indeed, what the tree consists of in the philosophy of idealism: *esse est percipi*—being is being perceived.

"Man is the measure of all things," declared Protagoras the Sophist,[2] "alike of the being of things that are and of the non-being of things that are not." Idealists like Bishop Berkeley, for whom no material objects exist apart from our consciousness of them, would agree with Protagoras. On the other hand, René Descartes, skeptic though he was, and John Locke viewed the world as consisting of real, palpable objects—particles, whorls, and vortices—with intrinsic primary qualities (such as hardness and roundness) and secondary qualities (such as color, taste, and sound), the latter of which are dependent on our sensations. Impatient with the seemingly ethereal constructions of their idealist opponents, realists are sometimes provoked to imitate Samuel Johnson's attempt to refute Berkeley by kicking a boulder to prove that it is *not* just a figment of our imagination. In this chapter I want to consider the problem of this boulder somewhat more closely, from the perspective of physical science.

What Classical Physics Regarded as Real

As it began its meteoric rise in the sixteenth century, science was hardly touched by the seemingly esoteric philosophical dispute about reality. The sun, the planets, the moon, falling stones, and colliding balls were the principal objects of interest. Doubts were raised, it is true, concerning the reality of the discoveries Galileo made with his telescope—some philosophers thought that only naked, unaided sense data were legitimate sources of knowledge.[3] But the entities in Newton's mechanics, such as forces, manifested themselves directly and powerfully to us, and any thought that reality might all be of our own making could safely be set aside. It was only Newton's pull of gravity, acting at a distance, that made people uneasy; Descartes's competing theory of vortex rings and swirling particles seemed much more palpable.

The physics of the nineteenth century, on the other hand, brought two basic changes that were harder for realists to accommodate. One was the atomic hypothesis that all matter consisted of particles. Though this idea had been introduced long ago by Democritus, it now became an important explanatory foundation of the nature of heat and the behavior of gases. All of statistical mechanics, the science underlying the laws of thermodynamics, was based on the hypothesis that gases, liquids, and solids were made up of atoms and molecules—Amedeo Avogadro managed to count them!—though no one had ever seen one of them, even under a microscope. The other change was brought about by Michael Faraday's introduction of the electromagnetic field, a concept even more intangible than molecules (see Figure 7). Originally envisaging the field in terms of invisible rubberband-like lines of force that connected particles carrying electric charge through free space, Faraday eventually abandoned the

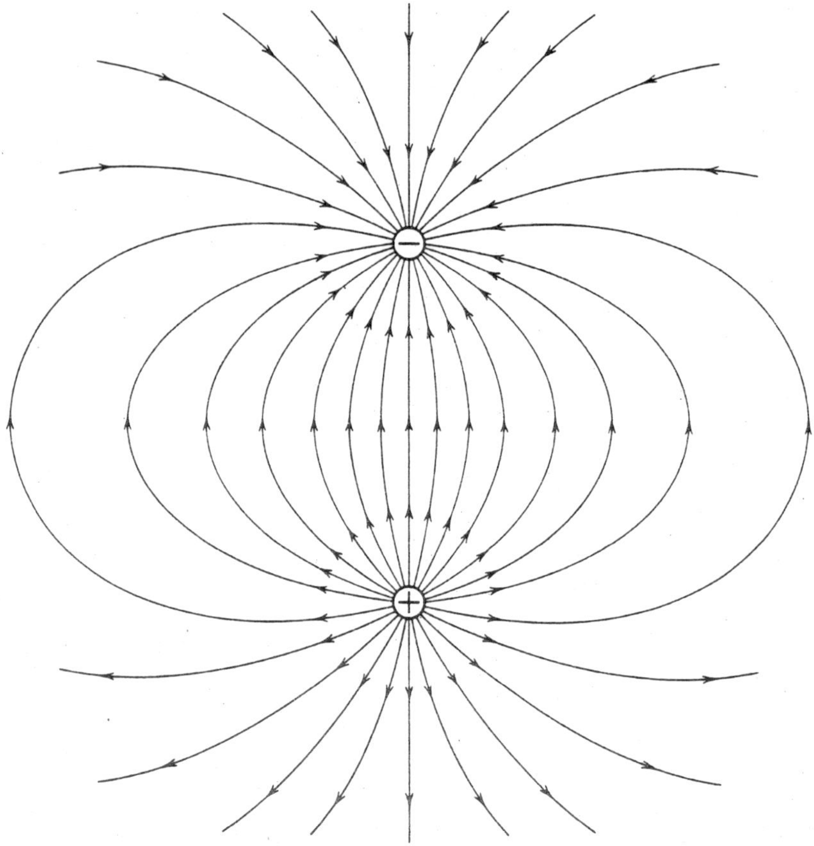

FIGURE 7 Electric field lines between a positive and a negative electric charge. (Reprinted from D. Halliday and R. Resnick, *Physics* [1986], part 2, p. 584, with kind permission from John Wiley & Sons, Inc., 605 Third Avenue, New York, NY 10158-0012.)

particles altogether and viewed them as no more than "centers of force: . . . the particle indeed is supposed to exist only by these forces, and where they are it is."[4] The change wrought by this new notion, whose mathematical formulation was subsequently provided by James Clerk Maxwell, was fundamental. "Before Maxwell," Einstein writes,

> the physically real—insofar as it was to represent what happens in nature—was imagined in terms of material points whose changes

> consisted only of motions . . . After Maxwell, the physically real was imagined in terms of continuous fields that were not mechanically interpretable . . . This change in our concepts of the real was the deepest and most fruitful in physics since Newton.[5]

The manner in which physicists viewed the field began to evolve, becoming ever more abstract. Maxwell still tried, unsuccessfully, to construct complicated mechanical models, and for some time it was still possible to use the ether as a crutch for the imagination—the forces might be seen as originating from a stress in its fabric—but the Michelson-Morley experiment destroyed any such easy way out. A further complicating factor has been the gradual enlargement of the idea of the electromagnetic field by the addition of many other fields, including gravitation and later the fields of the strong and weak nuclear forces, all of which began to be regarded simply as *conditions of space,* present everywhere, that determine the various forces exerted on different particles wherever they may be. Adding to the growing trend of assigning physical qualities to empty space, Einstein conceived of the force of gravity as a geometrical characteristic of the void; contrary to Kant's notion that the validity of Euclidean geometry for physical space is an *a priori* necessity, the geometry of space is non-Euclidean and depends on the distribution of matter. There is simply no way of avoiding the idea that free space is not featureless nothingness but is an entity that, like matter, requires study and observation to learn its many properties. If this seems strange and unintuitive, it is no more so than the way invisible atoms appeared at first to many scientists; the prominent and influential physicist Ernst Mach did not believe in the reality of atoms until shortly before his death in 1916.

Although similar questions also arise in psychology and the biomedical sciences (about, for example, the reality of mental states and disease), physics is the science in which the issue of the "real world" has become most pressing. In the early part of this century, the entities physicists regarded as existing—in addition to space and time—were fields (or waves) and particles like atoms and molecules, all of them real and "out there," not products of our making and imagining. It is clear from his own writing and the testimony of his associates that Ernest Rutherford—who discovered that the atom itself is almost entirely empty, with its tiny nucleus at the center surrounded by distant electrons—could see and feel atoms and electrons in his mind as palpably as if he held them in his hands. Many other physicists could claim a similar disposition: "I picture, without effort, the de Sitter space [in cosmology] as a four-dimensional surface in a five-dimensional space," Dirac once confided.[6] The more time and effort scientists spend working with even the most intangible objects, the more real and intuitive they become to them. "For my part," the philosopher Ian Hacking relates, "I never thought twice about scientific realism until a friend told me about an ongoing experiment to detect the existence of fractional electric charges [quarks] . . . '[In order to change the charge on a test ball of niobium,] we spray it with positrons to increase the charge or with electrons to decrease the charge.' From that day forth I've been a scientific realist. *So far as I am concerned, if you can spray them then they are real.*"[7]

If the atomic theory of matter and the concept of the field led some people to question what should properly be called "real," the same issue arose even more strongly with the new physics—relativity and the quantum theory—born early in this century. One of the consequences of Einstein's special theory of relativity was that the length of an object is contracted when viewed from a frame of reference in which it is moving. This effect had previously been called the Lorentz-Fitzgerald contraction, after its original theoretical discoverers. Their interpretation, however, differed from Einstein's: the object had, for them, a "real" length that shrank as it moved in absolute space or through the ether, whereas for Einstein, the effect was a simple consequence of the relativistic transformation law that relates lengths and times as measured in two frames of reference. If you are moving with respect to me, I see your clock running more slowly than mine and your yardstick looks shorter than mine, while you see my clock running slow and my yardstick shortened, entirely symmetrically.[8] This may raise a question in our mind about which clock is "really" slow and which stick has "really" shrunk; but our sense of what is real will simply have to adjust itself to the well-established experimental facts.

Enter the Quantum Theory

The emergence of quantum mechanics in the mid-1920s brought the question of reality most acutely to the fore. No discussion of what is *real* can avoid the quantum, so I will have to make a sizable digression and explain a number of aspects of quantum theory. This is an enormously successful paradigm, and although physicists have no difficulty in using and applying it, indeed building it into their working intuition, even some of its greatest practitioners and contributers admit they do not fully understand it.

One of Heisenberg's first announcements about his new "matrix mechanics" was his famous principle of indeterminacy, mentioned earlier. This principle forbids the simultaneous measurement of the position and the momentum of a particle, or, more generally, of any two specifically related so-called conjugate properties, with unlimited accuracy—the product of the two errors or uncertainties attending their measurement must always exceed Planck's constant. If we insist on determining one of them with great precision, we can be sure of the other only very roughly. The implication of this stricture is, of course, that, since knowing both the position and the momentum of a particle is necessary in order to calculate its future trajectory by means of Newton's laws of motion, we can never predict its behavior exactly. In fact, it makes no sense even to talk about the precise path of a quantum particle or about its motion and locations between successive observations.

More than that, the indeterminacy principle raises the question, does the particle *have* a precise location and momentum, which we are just not able to observe, or does it have no location and momentum at all? The answer given by Bohr and Heisenberg—generally (though not universally) accepted—is that a quantum particle *does not have* a well-defined position and momentum at the same time, though either one of them alone can be pinned down by a measurement as precisely as you wish; moreover, assigning it any such properties between observations has no meaning. From the time this doctrine emerged in 1926, after intense discussions in Copenhagen among Bohr, Schrödinger, Heisenberg, and others, it has become known as the *Copenhagen interpretation.*

The quantum theory takes an additional step that is destructive of traditional notions: it postulates that any two particles of the same kind, such as electrons, are literally *indistinguishable*—there is, as a matter of fundamental principle, no way of identifying one electron and telling it apart from another. (This indistinguishability is inextricably linked to the nonexistence of a well-defined trajectory. An infant learns that objects have a permanent identity by following their movements visually and recognizing their reappearance when they emerge from a concealed position. The movements of a quantum particle, however, cannot be followed—we cannot keep track of it or play peekaboo.) And this basic indistinguishability has important experimentally verifiable consequences in the statistical behavior of large collections of them: whereas classical particles with an individual identity follow the statistical laws promulgated by Maxwell and Boltzmann, quantum particles come in two varieties, fermions, which are subject to the statistics found by Enrico Fermi and Dirac, and bosons, which follow statistics discovered by Satyendranath Bose and Einstein. Electrons, for example, are fermions, photons are bosons. The difference in statistics arises from the fact that fermions obey Wolfgang Pauli's *exclusion principle*—no two of them can ever be found in the same state (this principle is at the basis of the periodic table of the elements)—whereas bosons do not. Clearly, in the quantum theory, a particle is not just a very small speck of dust.

Wave-Particle Duality

The year after Heisenberg announced his new matrix mechanics to account for the long-standing quantum puzzles with a full-fledged theory, Erwin Schrödinger published an apparently totally different theory, wave mechanics, with the same aim. His starting point was formed by a strange piece of that puzzle: whereas twenty years earlier, Einstein had introduced the concept of light quanta—particles later called "photons"—for what had been well established to be a wave phenomenon, Louis de Broglie had recently put forth the revolutionary idea that the microscopic entities known to physicists as particles—at the time only electrons and protons—also had a wave side to their nature. This wave aspect, Schrödinger proposed, was governed by his new equation. It shortly turned out that these two seemingly entirely different theories by Heisenberg and Schrödinger were, in fact, the same theory in different guises. If the first made an electron out to be fundamentally different from a tiny billiard ball, the second seemed to dissolve it completely into ethereal waves.

At the submicroscopic level, now, *everything* acquired a dual nature—both corpuscular and wave-like. Whereas formerly there was a fundamental distinction between wave phenomena (like light and radio waves) on the one hand and particle phenomena (like electrons and protons) on the other, all these physical entities now became both particles *and* waves. The force of the century-old interference experiments of Augustin Fresnel and Thomas Young, which had established that light must be a wave, could not be denied, but neither could the photoelectric effect, which could be explained only by Einstein's light quanta. The results of J. J. Thomson and Robert Millikan had unequivocally shown electrons to be particles, but those of Clinton J. Davisson and Lester H. Germer, which echoed Young's interference experiments for light, could be understood only if electrons, as de Broglie had predicted, were waves. There was no avoiding the conclusion that, at the micro level, Nature exhibits an ineluctable *wave-particle duality.* What is more, this duality of every entity has a chameleon-like quality: if we try to pin down its particle aspect, the wave disappears from view; if we make its wave side prominent and observable, its particulate nature dissolves.

The stripes observed on a distant screen when a beam of light is sent through two closely spaced slits in a wall are a concrete example. We do not see two luminous lines, simple replicas of the two slits. Instead, we see a whole series of bright and dark fringes, which are caused by the successive constructive and destructive interferences of the light emerging from the two openings, demonstrating unambiguously its wave nature (Figure 8). As the light intensity is dimmed, the image turns into a succession of flashes at sharp points, apparently randomly placed—light exhibits itself in the form of photons and no interference effects are noticeable. If the screen is replaced by a photographic plate, however, after long exposure a picture emerges showing the same light and dark lines that had been visible at higher intensity. The photons did not, after all, arrive at random places but were distributed in the striped interference pattern as before.

Well, you may say, if light consists of particles—photons—we ought to be able to find out through which slit a given photon traveled. So let us close first the left slit and then the right one. When all the photons have to come through the right one, of course, we obtain a picture different from the two-slit pattern, and when they all come though the left slit the picture is still different. However, printing the two images together, as a photograph with a double exposure, does

FIGURE 8 Diffraction fringes produced by light shining throught two vertical slits. (Reprinted from Sear, Zemansky, and Young, *College Physics* [1980], p. 728, with kind permission from Addison-Wesley Publishing Co., One Jacob Way, Reading, MA 01867.)

not give the same picture as when both slits are open! The interference fringes characteristic of two open slits are missing. In other words, when we try to pin down the particle aspect of light, assigning photons an identifiable path, we automatically destroy its wave nature. When we try, on the other hand, to be sure that it is a wave by observing the interference fringes with both slits open, we necessarily destroy its particle quality because we would be forced to say that each photon went through *both* slits. Remarkably enough, a similar experiment also works if the light beam is replaced by a beam of electrons. Both light and electrons have the same duality—particulate as well as wavy. Furthermore, which side of their nature they exhibit depends on the question put to them.

It was left to the philosopher in Niels Bohr to enshrine both the wave-particle duality and Heisenberg's indeterminacy in a vast new *principle of complementarity,* according to which everything under the sun has a dual nature with complementary facets that cannot be simultaneously viewed. Heisenberg's conjugate properties of a physical system are one instance; the wave-particle duality is another; truth and clarity, a third; perhaps life and its biochemical explanation, a fourth; and so on. Despite the enormous admiration Bohr universally enjoyed among scientists, many physicists did not go along with him on his journeys into mysticism; others, however, ventured even further into the swamp into which he had led them and lost their way.

What is a Particle?

Before exploring the perplexing questions about reality raised by the wave-particle duality inherent in the quantum theory, we should examine more closely what, exactly, we mean by "particles" in the ontology of basic physics, where that concept clearly looms very large. Heisenberg's version of quantum mechanics, in fact, was built entirely on the behavior of particles, and the quite different reformulation of the theory in the 1940s by Richard Feynman went to extremes in replacing all remnants of wave fields, even in electrodynamics, by the quantum motion of particles, forward and backward in time. All explanations and calculations in this theory are given in terms of the infinitely many paths that these particles might be imagined to have taken, and portions of his vision—the Feynman diagrams mentioned in Chapter 7—have become firmly entrenched in the imagination and language of physicists. Murray Gell-Mann and others have built this particle-based theory into a "histories" version that tries to avoid the Copenhagen school,[9] but these formulations and interpretations have, so far, not generally caught on, and their success at this point seems doubtful. In any event, their use of "virtual" histories of events raises at least as many reality questions as does the traditional formulation.

We clearly have to look more carefully at the basis of our understanding of particles. By this I do not mean such cornerstones of our view of the particulate nature of matter as Dalton's law of definite proportions in chemistry (chemical compounds are formed from their constituent elements in specific proportions), Avogadro's determination of the number of molecules in a volume of gas, and Einstein's explanation of Brownian motion. But we must ask on what grounds we regard the many "elementary particles" now known as particles.

Evidence for the first particle, the electron, came from two sources: J. J. Thomson's demonstration, by means of magnetic deflection, that the constituents of cathode rays had a definite ratio of electric charge to mass, and Millikan's measurement of the discrete nature of electric charges, always found to be whole-number multiples of an elementary unit, the "charge of the electron." The particulate nature of light was first indicated by Planck's derivation of the frequency distribution in "blackbody radiation"—the temperature-dependent electromagnetic radiation emitted by a black object—for which he was forced to assume that the energy of the emitted waves came in discrete packets proportional to their frequency. Perplexed as he was by the need for this mathematical device, Planck was too conservative to draw any revolutionary conclusions. It was Einstein who deliberately and consciously introduced the idea of "light quanta," or *photons,* as the only way to explain the mysterious results Philipp Lenard had obtained, called the "photoelectric effect": shining light on a metal surface caused electrons to be emitted, and their energies depended only on the color of the light while their number was determined by its brightness. The wave theory of light was powerless to explain this, but photons—particles whose energy was, *à la* Planck, proportional to the frequency of the light—made it easy: each electron was liberated by a single photon, which disappeared in the process, giving up its energy to that one electron.

It is not necessary to go through the experimental evidence for all the particles that now populate the zoo of what is known as "particle physics," but I want to mention certain noteworthy instances and general characteristics. The neutrino constituted an unusual case, because Pauli postulated its existence simply to account for the energy and angular momentum discrepancies in the radioactive decay of nuclei. (In desperation, Bohr had already been ready to give up the conservation laws.) The explanation Pauli offered struck many at first as *ad hoc,* like Popper's conventionalist stratagem, and of no scientific value, but the reality of neutrinos—massless particles interacting so weakly with matter that, as we now know, enormous streams of them constantly pass through the entire Earth unscathed—could not be denied when some twenty-five years later, "inverse beta decay" was detected: neutrinos produced in the decay reaction of one group of nuclei struck other nuclei and produced the inverse reaction. Neutrinos now play an increasingly important role in astronomy, though their detection still presents great experimental difficulties.

The half-dozen huge accelerators in high-energy laboratories built during the last forty years in various parts of the United States and Europe for the purpose of finding new particles have been very successful and led to an unexpectedly large number of discoveries. Produced in collisions among such atomic constituents as electrons and protons, the new particles are all unstable, meaning that they decay after extremely short periods of time into other particles. We naturally wonder how we know about, and indeed what we mean by, the existence of entities that "live" for only 10^{-15} sec (one millionth of a billionth of a second). If they live long enough so that, traveling at close to the speed of light (because they are produced with large kinetic energy in explosive collisions), they traverse a measurable distance, and there is no problem inferring their lifetime. In many instances, however, the lifetime is so short that the traversed path is not nearly long enough to be detected and the particle's existence has to be deduced in the following, more indirect maner. (Recall that in Chapter 5 I alluded to these inferences as examples of how theory-dependent such facts as the masses and lifetimes of unstable particles are.)

When particles are accelerated to high energies and made to collide with and scatter one another, they are finally caught in an array of detectors that measure what fraction was scattered and in what direction. This fraction, called the scattering *cross section,* varies with the energy with which the particles collide. Suppose the force between two particles is such that when their kinetic energy is near a particular value they are likely to be caught circling around each other for a long time before finally escaping from their dance. The existence of such a relatively long-lived state will lead to a large likelihood of scattering, that is, a large cross section, even if that temporary union cannot be thought of as really consisting of the two original particles but should much more plausibly be regarded as a new entity in its own right. In other words, *the evidence for the existence of some kind of unstable system or the formation of a "particle" in a collision is a sharp increase in the measured cross section;* the plot showing the variation of the cross section as a function of the kinetic energy of the colliding particles then contains a big bump, or "resonance," and the energy E at which this peak is centered is connected by Einstein's $E = mc^2$ to the mass m of the "particle" produced (c is the speed of light). What's more, the energy-time version of Heisenberg's indeterminacy relation gives a simple inverse relationship between the average length of time T for which the system exists and the width W

of the bump, that is, the "uncertainty" of its energy—the product of W and T must equal Planck's constant.

This is what high-energy experimenters mean when they announce the existence of a new unstable particle of mass m with a lifetime of T seconds: they see a sharp resonance peak in the cross-section plot centered at the energy corresponding to the mass m, whose width is equal to Planck's constant divided by T. The existence of the particle means just that and no more, except for the all-important fitting of this temporary existence into the whole framework of an underlying theory. This theoretical interpretation is usually of a very special nature: if certain features of the theory were slightly modified, say, by imagining that a particular interaction force has been "turned off," the unstable particle would become stable; its decay would be prevented for lack of an open physical exit mechanism. The "spikier" the observed bump in the curve is, the longer the lifetime of the unstable entity and the smaller the change in the forces necessary to make it stable. On the other hand, a wide bump—a very gradual rise and fall is difficult to identify and ambiguous in any case—indicates a shorter lifetime of the hypothetical unstable system and a larger interaction that would have to be ignored in order to transform this system into a stable one, at the same time rendering the appellation "particle" less convincing. Thus the borderline between what is called a particle and what is not worthy of that name is, in principle, quite vague.

To appreciate just how blurred this distinction is, you have to realize that every experimental datum is subject to possible error—the particles in the beam do not all have exactly the same energy; the position of a detector is not precisely defined because its size isn't only a point, nor are the counters registering the arrival of particles 100 percent reliable, etc. Consequently, repetitions of the same experiment will generally not yield exactly the same results. For this reason, when experimenters draw a plot of their data, such as the observed scattering, they show the outcome of a statistical analysis of many individual detector counts by indicating how much that particular number might be mistaken, attaching to each point a vertical bar—its length marking the probable range of error (see Figure 9). These bars are exactly analogous to the way newspaper reports of the latest political polls indicate the statistical range of errors; they

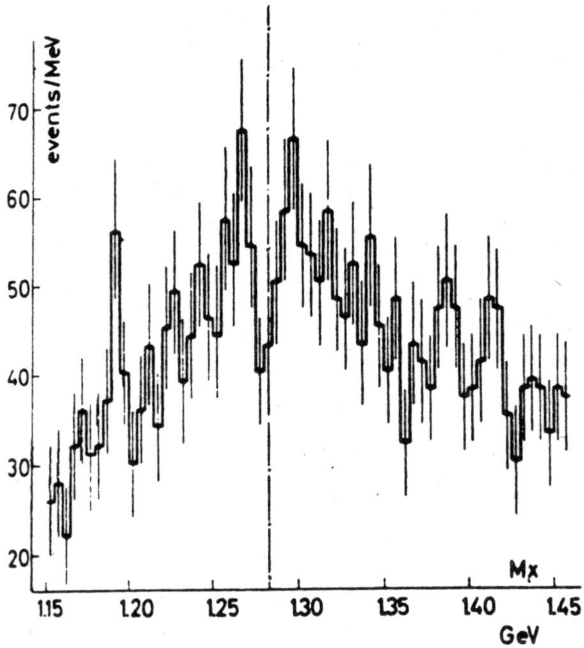

FIGURE 9 A plot of experimental data, including error bars. (Reprinted from G. Chikovani et al., *Physics Letters 25B* [1967], p. 47, with kind permission from Elsevier Science–NL, Sara Burgerhartstraat 25, 1055 KV Amsterdam, The Netherlands.)

are a measure of how much the reported numbers might differ in a repetition of the same poll and in a count taken on the entire population. Such statistical results can sometimes be very misleading, as the following controversy illustrates.

In 1967 a group of physicists at the large particle accelerator at CERN, the European high-energy physics laboratory in Geneva, Switzerland, published their discovery of a resonance, then called the A_2, visible in the plot of collision cross sections of protons as functions of the energy. The A_2 had a very unusual structure, consisting of a double bump. Its appearance of being split in two suggested the existence of *two* particles of almost the same mass. Another group from Northeastern University, working on a very similar experiment, failed to see the dip in the bump, in spite of strenuous attempts to find it in their data. Perhaps their apparatus was not sensitive enough to detect such a fine structure in the curve? The issue came to a head at a meeting of the American Physical Society in 1971, at which the CERN group argued that since they could see the split while the other group could not, their instruments were obviously more sensitive. (Recall the discussion of the sensitivity of experimental equipment in Chapter 2!) As more data accumulated, both groups eventually agreed that the splitting of the A_2 was a statistical artifact, and the issue disappeared.[10]

You might think that the simplest alternative to such an imprecise definition is to reserve the name "particle" for stable objects only. This, however, would have grave drawbacks of its own. We all know that atomic nuclei are made up of protons and neutrons, and, as the continuing solidity of the chair on which I am sitting indicates, most of these nuclei are stable. The neutron, however, when out in the open, is unstable, with a half-life of some 13 minutes. (Inside a nucleus it is stable only because Pauli's exclusion principle prevents the proton among its decay products from finding an available state.) All the evidence we have, on the other hand, indicates that the proton is stable, though this stability is at the present time subject to very rigorous testing. Would we, then, want to call the proton a particle and deny that title to the neutron, all of whose properties, save its electric charge, are extremely similar to those of the proton? In any case, this would merely be a linguistic solution—the unstable entities, whatever we call them, still have many particle-like features.

And then there are the *quarks,* the particles now thought to be the ultimate building blocks of all others. Stable they are, but they have another property that gives their reality a strange cast—they have never been, and present theory says they *can* never be, detected separately, outside other particles like protons or neutrons. Originally proposed by Gell-Mann as no more than a mathematical device to account, in a systematic fashion, analogous to the periodic system of the elements, for the ever-proliferating number of particles found in collision experiments, they not only serve that function very efficiently but are now regarded as no less real than the many others. Quarks, indeed, are not the only particles whose existence is confined to the interior of other entities. The quantum theory of matter makes extensive use of *phonons,* which play the same role for the mechanical vibrations of atoms and molecules in solids and liquids as do photons for electromagnetic oscillations—phonons are for sound what photons are for light. Since they are vibration quanta, they can never be found outside solid or liquid materials, of course, just as quarks cannot exist outside other particles.

The entire particle concept, which in Newtonian physics and in the Cartesian view of the world was both fundamental and unproblematical, was simultaneously substantiated and dissolved by the quantum theory: on the one hand, this theory explains the stability and unvarying identity of atoms as the building blocks of all matter; on the other hand, the wave-particle duality makes the very concept of localization that lies at the heart of the notion of a point particle nebulous. That this is unavoidable is illustrated by considering the idea of brief light flashes. We have no difficulties imagining a blink of light—such signals play a fundamental role in Einstein's arguments leading to the theory of relativity—nor is it difficult to think of light as colored. What could be wrong with a momentary red flash? If the idea of a monochromatic light signal of very short duration is carried to an extreme, however, the classical Maxwellian theory of the nature of light—this has nothing to do with the quantum theory—tells us that the concepts of a single color and instant flash are contradictory; we cannot realize them both at once.[11] It is the fact that all matter, and not just light, has a wave nature that leads to the abandonment of the notion that point particles moving with a well-defined momentum can be localized precisely, and hence finally to the dissolution of the particle concept and the idea of localization. If ever you thought of particles as possessing material *substance,* you must clearly relinquish that notion.

Scale-Dependent Realism

Does all of this strike you as more and more "unreal"? Are these submicroscopic particles nothing but figments of our imagination, as Mach thought atoms were? If quarks are not real, why should neutrons be, and if neutrons are not, why atoms? Recent devices have made it possible to show "pictures" of individual atoms, but though they will surely become increasingly sensitive, exhibiting more and more detail, I doubt we will ever be able to "photograph" the quarks inside the protons in the interior of a nucleus at the center of an atom. Does that give them less reality than atoms?

Bohr's reaction to such ontological questions was unequivocal. Always professing a lack of interest in *reality,* he placed his emphasis on *language.* "What is it that we human beings ultimately depend on?" he asked.

> We depend on our words . . . Our task is to communicate experience and ideas to others. We must strive continually to extend the scope of our description, but in such a way that our messages do not thereby lose their objective and unambiguous character . . . We are suspended in language in such a way that we cannot say what is up and what is down. The word "reality" is also a word, a word which we must learn to use correctly.[12]

Accordingly, he came to the conclusion that

> There is no quantum world. There is only an abstract quantum mechanical description. It is wrong to think that the task of physics is to find out how Nature *is.* Physics concerns what we can *say* about Nature.[13]

Heisenberg had similar views, but from a somewhat different perspective. He, too, placed a strong emphasis on language: "Every description of phenomena, of experiments and their results, rests upon language as the only means of communication. The words of this language represent the concepts of classical physics . . . Therefore, any statement about what has 'actually happened' is a statement in terms of the classical concepts."[14] But he went further. Some find it desirable, he wrote,

> to return to the reality concept of classical physics or . . . to the ontology of materialism . . . This, however, is impossible . . . It cannot be our task to formulate wishes as to how the atomic phenomena should be; our task can only be to understand them.[15]

And later,

> In the experiments about atomic events we have to do with things and facts, with phenomena that are just as real as any phenomena in daily life. But the atoms or the elementary particles are not as real; they form a world of potentialities or possibilities rather than one of things or facts.[16]

As he saw it, "The ontology of materialism rested upon the illusion that the kind of existence, the direct 'actuality' of the world around us, can be extrapolated into the atomic range. This extrapolation is impossible, however."[17]

The essential point to be stressed, and to that extent I agree with both Bohr and Heisenberg, is that *realism is a matter of scale.* It is one thing to be a realist at the scale of everyday life and experience, but quite another to try to carry that realism to the micro world, where neither our experience nor our language is adequate. We insist upon formulating what is happening at the micro level in terms of either "particles" or "waves," and in order to understand it—not just mathematically—we seem to have no other choice. While the results of observations and experiments can and must be described in a "classical," everyday language, the micro phenomena are not suited to such a vocabulary.

The necessity of describing observations in the language of classical physics explains why the Copenhagen interpretation insists on placing the quantum theory in a classical framework, with which it must make contact in every measurement. Logically, Bohr insisted, "by the very word 'experiment,' we refer to a situation where we can tell others what we have done and what we have learned."[18] After quoting Carl Friedrich von Weizsäcker, "Nature is earlier than man, but man is earlier than natural science," Heisenberg then comments, "The first part of the sentence justifies classical physics, with its ideal of complete objectivity. The second part tells why we cannot escape the paradox of quantum theory, namely, the necessity of using classical concepts."[19]

Some physicists regard this encapsulating as very unnatural, arguing that the quantum theory ought to be able to stand on its own feet, with the laws of classical physics arising as "limiting cases" when, on the scale of everyday phenomena, Planck's constant may be considered negligibly small. This is what the formulations by Gell-Mann and Hartle and by Omnès, mentioned earlier, try to accomplish. There is, of course, no question that the classical laws must indeed, and in fact do, emerge in such a sense—though the demonstrations of why we need not worry about wave functions, quantum interference, and the like while observing the motion of the Moon, and why we see the tracks of electrons in photographic emulsions, are not simple—just as Newton's laws of motion represent the limits of Einstein's when all velocities are small relative to the speed of light. But it is also hard to deny that for the description of the outcome of experiments on micro systems governed by the quantum theory, we are forced to use the language of classical physics. The point is not that the observed quantum system is small and the observing apparatus large and therefore governed by classical laws—this is not always the case—but that there is no alternative to the use of everyday language for communicating the results of the observations.

So much, then, for a tour through the province of basic physics populated by waves and particles. The sights on that tour should have convinced you that the concept of "particle," convenient as it is for many purposes and unavoidable as it seems to be for an account of what is real on one scale, is not suitable for conceptualizing reality at the most basic level. So we should now return to the more general problems arising from the quantum theory, and especially from the wave part of the fundamental wave-particle duality, for it is in this area that the most puzzling questions concerning the description of "reality" arise.

REALITY AT THE SUBMICROSCOPIC LEVEL

THE question of how to deal with reality at the submicroscopic level has been around ever since the Greek philosophers began to imagine the material world to be made up of unseen solid entities like tiny atoms. As our observational tools have become increasingly refined and powerful, this question, far from having been answered, has grown more and more puzzling; it remains the task of physical science to answer it as best it can. We therefore have no choice but to delve more deeply into the strange theory designed to understand submicroscopic matter—quantum mechanics. Once we have done so we shall sail into smoother waters again in the last chapter, but in the meanwhile some readers may find the voyage a little rough, demanding both patience and close attention. Remember that my aim is not just to exhibit, for its own sake, a specific theory of physics, with all its quirks and special concepts, but to attain the closest view of our comprehension of reality at the submicroscopic level that we have at the present time.

In Schrödinger's formulation of quantum mechanics—the one principally used by physicists in practice—the primary descriptive tool for the behavior of a well-specified physical system is the wave function, the mathematical expression of the *state* of the system, as discussed in Chapter 8. But what is the concrete physical meaning of this function? Is it a "condition of space" like the electric field? Is it *real?* Schrödinger himself first regarded the wave function as analogous to a field, but he came to realize that such an interpretation could not work: an electric field has a specific value at a given point in three-dimensional physical space, but the wave function "lives" in what is called *configuration space*—that of two electrons in a six-dimensional space, that of three electrons in a nine-dimensional space, etc.[1] The meaning now universally agreed upon originated with Niels Bohr, Max Born, and Pasqual Jordan: the wave function defines a *probability*—the square of the wave function of an electron at a given point in space is the probability of finding the particle at that point.[2] On the basis of my earlier discussion of the concept of probability, you may guess that this definition, though unambiguous from a practical point of view, is conceptually a can of very squiggly worms.

The probability definition of the wave function has given rise to a variety of interpretations, all originating from different views of the meaning of probability. There are, first of all, the subjective interpretations, according to which the wave function codifies an experimenter's *knowledge* of the system. Since the wave function embodies all that the quantum theory can assert about the state of a physical system, this interpretation automatically introduces an element of subjectivity into Nature at a very basic level, and though it has some very prominent adherents, most contemporary physicists, when pressed, would reject it in favor of an objective construal; but such a reading has its own problems and runs willy-nilly into the ambiguities attached to the interpretation of probability outlined in Chapter 8. Recall that the reason Popper introduced his propensity theory was to avoid having always to deal with ensembles and to be able to give the wave function an objective probability interpretation attached to *individual systems,* which physicists much prefer.

Collapse of the Wave Function

The difficulties concerning the interpretation of the wave function become particularly acute in the discussion of the process of observation or measurement. To understand why, consider a physical system in a state that is as well specified as possible, and therefore described quantum-mechanically by a wave function (also called a *wave packet*) whose change in time is governed by the Schrödinger equation. Even so, the parameters of the system—such as the position of a particle—are not precisely predictable but can be given only in terms of probabilities; in a sense, they are "smeared out." Suppose, now, we perform a fairly precise measurement of the position, thereby determining it within narrow limits; this localization drastically reduces the "smearing out" and the wave packet *collapses* everywhere. If the system is left alone from that point on, its new state is described by a "reduced wave function," which again develops according to the Schrödinger equation.[3]

Let us take a particularly simple example that will serve well for later purposes. Many particles have a property called *spin,* which you may think of as a quantum version of a gyroscope-like clockwise rotation about an axis through the center of the particle; the direction of the spin is the direction of this axis. According to the quantum theory, the directional components of the spin are "quantized," which means that if we measure its vertical component, the only possible results are integral (0, ± 1, ± 2, etc.) or half-integral ($\pm 1/2$, $\pm 3/2$, $\pm 5/2$, etc.) multiples of Planck's constant, with the maximal magnitude determined by the particle's intrinsic spin; similarly for its east-west and north-south horizontal components. In the case of electrons, the only possible measurement outcomes are $+1/2$ and $-1/2$ (in units of Planck's constant), or + and − for short, which you might also call *up* and *down* for the vertical component or *right* and *left* for a horizontal one. Such measurements are done by taking advantage of the particle's *magnetic moment,* which makes it act like a tiny bar magnet pointing in the direction of its spin. A device first used by Otto Stern and Walther Gerlach (which I shall call an *SG apparatus*), when oriented vertically, contains a magnetic field exerting an upward force on a magnet tilting up and a downward force on a magnet tilting down. The result of sending a beam of electrons through such a device is to split the beam into two, one directed at a definite angle upward, the other at a definite angle downward. This has to be interpreted as meaning the electrons in the upper beam have spin *up* and those in the other beam have spin *down.* (If all the electrons in the initial beam happen to have spin *up* or all spin *down,* of course, there will be only one deflected beam.) An SG apparatus oriented horizontally will analogously produce two horizontally deflected beams, one with electrons of spin *left* and one with electrons of spin *right.* The finding by Stern and Gerlach that, for a given orientation of the apparatus, there are no other angles of deflection but these two was a remarkable discovery—the quantization of the electron's spin angular momentum. (See Figure 10.)

The story begins with a beam of electrons, all with spin *up*. If we determine their vertical spin components by sending them through a vertical SG apparatus, they will be deflected upward, confirming that their spins are pointing *up*. Next we measure their horizontal spin components, by sending them through a horizontal SG device: we then find that the beam is split into two equal parts, one beam with spin *right* and the other with spin *left*. These results are in agreement with the quantum-mechanical prediction concerning the spin-*up* wave function of the initial electrons, that the probabilities for horizontal spin *right* and spin *left* are each 1/2; in accordance with Heisenberg's indeterminacy principle for the two spin components, the horizontal components are completely "smeared out" when the vertical spin component has a definite value. (Intuitively you may picture such an electron as a spinning top whose axis is precessing in a circle about a vertical line, giving it an equal chance of being found at any moment tilting to the right or to the left.) The two separate beams are the result of the measurement, which "reduced" or collapsed the original wave packet of spin *up*, the left beam being described by a spin-*left* wave function and the right beam by a spin-*right* one; we have definite, verifiable assurance that the spin of every electron in one beam points to the left, the spin of those in the other beam to the right. If we guide the two beams together, making one mixed beam out of the two, the resulting state is not as well specified as that of the original beam: it is no longer all spin *up* but a mixture of half spin *left* and half spin *right*. In other words, the new beam is *not describable by a wave function*.[4] (A vertical SG measurement would yield half *up* and half *down*.)

There is also, however, a way—very difficult to accomplish in practice—of carefully recombining the two separate beams without disturbing either one. Such a reconstituted beam is now not a *mixture* of electrons whose horizontal spins are known but again a beam of electrons all of which are of spin *up*. The result is sometimes expressed, rather misleadingly, by saying "the outcome of the horizontal spin measurement was not recorded"; the first-mentioned method of recombining the beams with less care, in contrast, is expressed as

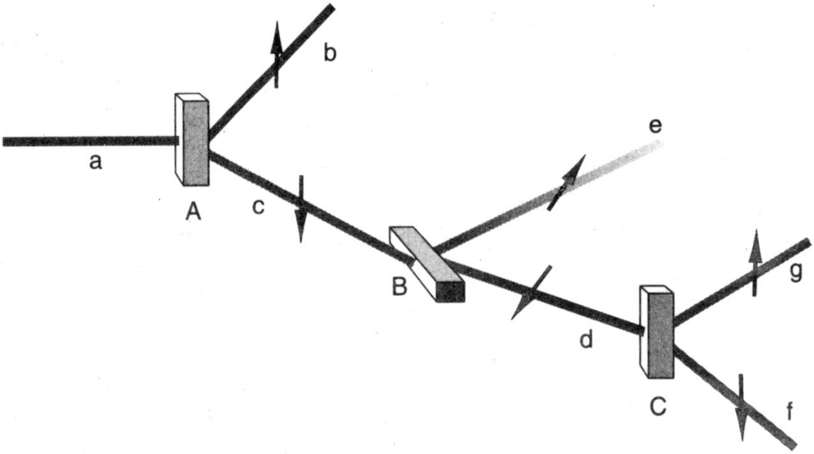

FIGURE 10 The SG apparatus A tests the particles in beam *a* for spin up or down; those with spin up go along path *b*, and those with spin down follow *c*. Instrument B tests those particles in beam *c* for spin left or right, and those with spin right follow path *d*, while those with spin left follow *e*. Apparatus C again tests the particles with spin right for spin up or down, and half are found with spin down, following *f*, while half have spin up and go along *g*.

"the outcome of the measurement was recorded." Some physicists have even suggested that what is crucial is whether the record was registered in a human consciousness, an interpretation that cannot be justified. The important fact about the two separate beams after the original beam was split by the horizontal SG device is that they are still *entangled*.[5] Their entanglement is broken by any attempt to extract, from either one, an electron, whose horizontal spin would then be definitely determined. *This* is what would make the act of splitting the beam an actual measurement, leading to the destruction of the entanglement and to the reduction of the wave packet.

The collapse of the wave function caused by a measurement, and by any observation as well, has led to an enormous amount of discussion. It is particularly disturbing to people who intuitively (but mistakenly) interpret the wave function as a condition of space, because it appears to be an instantaneous effect in the entire universe. In fact, however, the so-called collapse—which appears on the surface to be a strictly quantum-theoretical phenomenon, as Popper[6] pointed out—is a necessary characteristic of any probabilistic theory.

On the other hand, the property of entanglement—which is not at all odd for waves, extended in space as they are, but which seems strange for particles—implies some very unintuitive results at the very heart of the reality question, as a great debate between Einstein and Bohr demonstrated. It is, at bottom, the wave-particle duality that leads to this entanglement, owing to the fact that the joint probability of two events is not always simply the product of the two separate probabilities.

The probability that two players, each with their own deck, who keep turning over cards will both turn over aces in the next round of play is the product of the two probabilities that each of them will do so alone—the two events are independent. But the probability for one player to turn over two aces in a row from the same deck is not the product of the single probabilities—the two events are not independent, they are entangled. Entanglement of particle-events that are spatially far separated, however, is counter-intuitive and strikes us as weird, because our intuitive grasp of corpuscles is that they are *individual* and *localized.* We have no instinctive feeling for mutual dependency of bounded objects, while we have little difficulty understanding the interdependency of extended entities like waves, which may overlap.

Schrödinger's Cat

The question of the meaning of the wave function and of its collapse at the instant of measurement is brought into sharp relief by a conundrum posed by Schrödinger, who was not happy about the generally accepted Bohr-Born-Jordan interpretation of his creation. Schrödinger imagined a cat imprisoned in a cage together with a diabolical device consisting of a single radioactive atom with a half-life of an hour (see Figure 11). The atom's decay, signaled by the click of a Geiger counter, would cause a bottle filled with a deadly acid to break, killing the cat. Schrödinger imagined the wave function of the entire system—cat and atom—an hour later, when there would be a fifty-fifty chance that the atom had decayed: if it had decayed, the bottle would be shattered and the poor cat would be dead; if it had not, the cat would be alive and happy. Before we open the door of the cage to check, it seems, the cat is half alive and half dead, because this is what the wave function of the atom-cat system says; only after

FIGURE 11 Schrödinger's cat in an airtight cage, with a hammer that smashes a vial of poison gas when triggered by the radioactive decay of an atomic nucleus. (Reprinted from J. R. Brown, *The Laboratory of the Mind* [1991], p. 24, with kind permission from Routledge, 11 New Fetter Lane, London EC4P 4EE, U.K.)

opening the door for inspection is the cat definitely alive or definitely dead, or so it appears. Schrödinger concluded that, since this proposition was absurd, the customary interpretation had to be wrong.

A group of physicists at the National Institute of Standards and Technology recently tested an analogue of this fanciful construction in an actual laboratory experiment.[7] Using very sophisticated laser technology, they were able to trap a single beryllium ion, whose spin could point either up or down, in such a way that it was in one position if its spin was *up* and in another, far distant (by atomic standards) position, if it was *down*. Indeed, they managed to produce a state described by a wave function of an ion *here* with spin *up* and at the same time *there* with spin *down*—like the cat alive with an undecayed atom and the cat dead with an atom decayed. Only in response to an inspection in order to ascertain its position would the beryllium ion "make up its mind" to be either in one place or the other.

After our first shock at the strangeness of all this has worn off, it becomes clear that the situation described is, of course, no different from the simple wave function of a spin state half *up* and half *down*. What offends our intuition in the case of Schrödinger's cat is that it is one thing to get used to a quantum property like spin not having a definite value like up or down, but, by God, whether we open the cage or not, a cat is either dead or alive, and an ion is either here or there! Is the quantum theory really saying that the Moon isn't there when no one is looking at it?

It says no such thing. The confusion arises from the fact, which I emphasized before, that the *state of a physical system*, as described by the quantum-mechanical wave function, does not mean the same as it does in classical physics and in our intuition based on everyday experience. The wave function of the cat-and-atom system[8] describes its state in terms of probabilities, either in the sense of describing only an ensemble of such systems or else in the sense of Popper's propensities—its *quantum state;* its *classical state,* for which we have some intuition, is something else. The woman whose husband has gone to a gambling casino might well say that he is in a state (in the quantum sense) of rich-poor until he comes home and shows her his wallet. Ultimately, the befuddlement arises from the mistaken notion that a quantum-theoretical state, as described in the ideal case by a wave function, is a direct description of reality.

The EPR Debate

In 1935, Einstein and two young collaborators, Boris Podolsky and Nathan Rosen, published a paper in the *Physical Review* that presented a detailed argument for a negative reply to the question asked in its title, "Can quantum mechanical description of physical reality be considered complete?" The paper was answered within a few months by Bohr, in an article with the same title in the same journal. The EPR paper, as it is usually referred to, gave rise to a large body of commentary by philosophers and scientists, continuing to this day, since it goes to the heart of what most bothered Einstein and others about the quantum theory. Not only did Einstein reject the notion, we recall, that "God played dice with the world," but more important, he was "inclined to believe that the description of quantum mechanics . . . has to be regarded as an incomplete and indirect description of reality, to be replaced at some later date by a more complete and direct one."[9] The argument, in a nutshell, as later simplified by David Bohm, runs as follows.

Assume that a molecule with zero angular momentum, consisting of two atoms of spin 1/2, breaks apart with some excess energy, so that the two atoms fly off in opposite directions. By the law of conservation of angular momentum, the spins of the two daughter atoms must add up to zero, which implies that their spins must point in opposite directions—if the spin of one is *up,* the spin of the other must be *down,* no matter in what common direction the two SG measuring devices are oriented. If we send atom 1 through a vertical SG device and find its spin to be *down,* we can be sure, without ever getting near it, that the spin of atom 2, which may be far away, is *up.* (We could, of course, verify this by means of a vertical SG measurement on atom 2, but there is no need for that.) According to the definition given by EPR, *"If, without in any way disturbing a system, we can predict with certainty . . . the value of a physical quantity, then there exists an element of physical reality corresponding to this physical quantity";*[10] the vertical spin of atom 2 therefore corresponds to "an element of physical reality."

The same conclusion holds for a horizontal spin measurement—we could have made the measurement on atom 1 by orienting the SG apparatus horizontally. Therefore, the horizontal component of the spin of atom 2 also corresponds to "an element of physical reality." Because of Heisenberg's indeterminacy principle, however, quantum mechanics does not permit us to determine the vertical and horizontal spin components simultaneously. *Ergo:* "quantum mechanics has to be regarded as an incomplete description of reality."

In our earlier terminology, the spins of the two atoms are *entangled.* If, after subjecting atom 1 to a vertical spin measurement with the result *down,* we send atom 2 through a vertical SG device, we will always find its spin *up;* but if we do the same after making a horizontal spin measurement on atom 1, there will be a fifty-fifty chance for the spin of atom 2 to be *up* and a fifty-fifty chance for it to be *down.* How does atom 2 "know" what measurement was done on atom 1? We would have to explain what is going on by an instantaneous communication between the two—"spooky action at a distance," Einstein derisively called it. It should be noted that this imaginary experiment shows the interpretation of Heisenberg's indeterminacy principle as expressing an uncontrollable disturbance caused by a measurement to be untenable, unless that disturbance can be communicated instantaneously from one particle to another far away.

There you have it: so far as reality is concerned, absent "spooky action at a distance," EPR insists, quantum mechanics cannot be the whole story. And here is part of the reply from Bohr, to whom any disagreement with Einstein was agonizingly painful:

> The extent to which an unambiguous meaning can be attributed to such an expression as "physical reality" cannot of course be deduced from *a priori* philosophical conceptions, but . . . must be founded on a direct appeal to experiments and measurements . . . In fact, this new feature of natural philosophy means a radical revision of our attitude as regards physical reality.[11]

Pauli, who considered the search for an unknowable objective reality to be analogous to the medieval scholastic question of how many angels could dance on the point of a pin, agreed with Bohr, and so did Heisenberg.

The puzzling nature of the entanglement of the two particles in the EPR thought experiment, however, is simply a consequence of the probabilistic nature of the quantum theory, as the following model will show. Let us play a game in which a central player repeatedly throws pairs of identical balls, one to a catcher in left field, the other to a catcher in right field. The balls come in two slightly different sizes and two colors, green and red, the properties of size and color being independent of each other, and they are thrown randomly, that is, with the same probability of 1/4 for each of the four possible combinations, red-big, red-small, green-big, and green-small. Each catcher is allowed to pay attention to only one property—either the color or the size of the ball she catches. If catcher 1 finds that her ball is red, she knows that the ball thrown to catcher 2 is also red, and if she finds that her ball is small, then so is the other. Now, if catcher 2 tests the size of her ball after catcher 1 has found hers to be large, she is certain to find hers large too; but if she tests its size after catcher 1 has found hers to be red, her chance of finding hers to be small is fifty-fifty. We are tempted to ask the same question here as we asked of the pair of atoms: how does the ball caught by catcher 2 "know" what test was performed on the other ball?

This model game is not intended in any sense as a "realistic" explanation of the EPR experiment; after all, here, the EPR criticism is seen explicitly to be correct—the description of "reality" by each player is deliberately incomplete, since each is allowed to test only one of the two properties of the balls they caught. The only purpose of the model is to demonstrate that the puzzling "entanglement" of particles is an effect of the probabilistic character of the quantum theory. In order to achieve entanglement classically, the description *must* be incomplete—each player may pay attention only to one of the two properties of a ball—whereas in quantum theory even a complete description can lead to it. This is the crucial difference between classical and quantum physics.[12]

The physicist David Bohm[13] and his followers have been trying for a long time to account for the quantum probabilities by introducing a deterministic substratum analogous to the way the classical molecular theory of matter underlies statistical mechanics. This substratum is meant to consist of particles that are themselves in principle unobservable, their sole function being to make the theory more intuitively acceptable by grounding the probabilities of quantum mechanics on our ignorance of more fundamental processes never accessible to observation. Thus, in agreement with the conclusion of EPR, entanglement would be the result of an incomplete description. There are those who argue[14] that but for historical contingencies, Bohm's interpretation rather than the Copenhagen version might have become dominant among physicists. This may be so, but the quantum world would not have been stripped of its strangeness; only its description would have been different: the Bohm theory cannot account for the probabilistic nature of quantum processes without introducing, at the most basic level, intrinsically nonlocal effects that are the precise analogues of Einstein's "spooky action at a distance." The laws governing the substratum, in other words, *cannot be* comfortably intuitive and classical. In comparison with the orthodox theory, the drawback of Bohm's is that, for the sake of some partial increase in intuitive appeal, it postulates *ad hoc* an unobservable netherworld, thereby violating the rule of Ockham's razor. Some form of the quantum puzzle cannot be avoided.

Bell's Inequality

For thirty years the issue raised by the EPR paper remained entirely on the philosophical level, debated but inconclusive. Then, after Einstein's death, the Irish physicist John Stewart Bell shifted the ground to a question that was, in principle, amenable to experimental testing. He devised a numerical inequality that, for certain phenomena, had to be satisfied by any "realistic" and "local" theory—with no influences traveling faster than the speed of light—but was definitely violated by a quantum-mechanical explanation. Let me give you an illustration of Bell's inequality in a special case.[15]

The setup consists of a transmitter at the center and two receivers at some distance on either side of it. Each of the receivers, between which there is no direct communication, has a dial with three settings and a light that can flash either green or red. (See Figure 12.) One round of the game to be played proceeds as follows: after the dials at the two receivers have been set, independently, unbeknown to one another and to the transmitter, the latter simultaneously sends identical signals to the two receivers, and in response, each of the receivers flashes its light in one of its colors, red or green. The game, whose results are recorded, consists of many rounds, with randomly varying signals and different dial settings.

Now imagine this game played using quantum particles with spin 1/2 for signals, as in the EPR setup described, and with receivers equipped with SG devices oriented at special angles determined by the dial settings. The rules of quantum mechanics lead to the simple statistical conclusion that if the dial settings are ignored in the count, *the colors flashed by the receivers agree exactly half the time.*

The question now is how to play this game and achieve the same statistical result by means of *realistic* signals, with messages that are

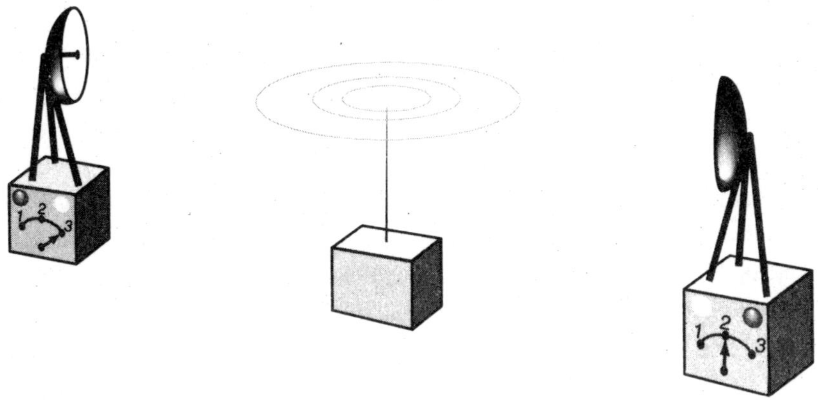

FIGURE 12 The setup for an EPR experiment: A signal emitter and two receivers with dials and colored lights.

"elements of reality," telling each receiver which color to flash, depending on its dial setting—say, by explicit instructions sent in some code. Among the $2 \times 2 \times 2 = 8$ different possible instructions sent to the receivers (two possible colors for each of the three dial settings), there are two that always result in equal colors of the two lights, no matter the dial settings: those that say "green for all settings" or "red for all settings." Since there are $3 \times 3 = 9$ different pairs of dial settings, the result will be 18 instances of equal colors. I will leave it to the reader to verify that each of the other 6 possible instructions will result in 5 cases of equal colors and 4 of unequal ones. That adds up to a total of $18 + (6 \times 5) = 48$ instances of equal colors out of a total of $8 \times 9 = 72$, or a probability of $48/72 = 2/3$. Thus, any realistic method of signaling will yield a probability for equal colors that is definitely greater than 1/2, a special case of Bell's inequality; quantum mechanics, however, violates this inequality by giving the result 1/2. (Of course, we could also get the result 1/2, or even less, by allowing the two receivers to be in instant communication with one another when the signals arrive and allowing each to adjust their settings depending on that of the other—this is what Einstein regarded as "spooky.")

Bell's theorem represents a clear-cut, experimentally verifiable procedure to decide whether there might be a more complete realistic theory to account for the phenomena, as Einstein was hoping, or whether, realism or no, the quantum theory is the best we can do. A number of experimental tests using Bell's inequality did, in fact, take place at various laboratories—thought experiments were transformed into real ones—and although the results always favored quantum mechanics, they are still somewhat controversial. Nevertheless, it looks as though there is no hope for replacing the quantum theory by one that satisfies our hunger for a realistic description of the micro world in everyday language and also shuns "spooky action at a distance."

The Crucial Importance of Quantum Field Theory

Many of the weird and counter-intuitive results of the quantum theory arise simply from our irrepressible urge to describe the world in terms of either particles or waves. In a language that describes reality at the submicroscopic level without these notions, however, many of the apparent paradoxes can be expected to disappear. Such a language does exist: in place of the familiar particles and waves, it uses the concepts of the quantum theory of fields. To explain some of its rudiments, I have to make a side trip into the land of quantum electrodynamics, or QED. QED is the original creation of Dirac, who conceived it a few years after Heisenberg's and Schrödinger's new quantum mechanics. Making this detour will also bring us a little closer to an understanding of how particles—photons, the "particles" of light, as well as electrons—can arise from the quantum theory without being put into it from the start.

Recall that the idea underlying quantum mechanics was to replace the numerical quantities that specify the classical state of a system of particles (for example, their positions and momenta) with "operators"—mathematical objects that generally change a given function into a function of a different form. Operators have the characteristic property of not necessarily "commuting" with one another. In other words, the product of operators a and b depends on their order: $a \times b$ need not necessarily be equal to $b \times a$. This replacement of the classical, numerical dynamical variables by operators with specific commutation properties—specifying by how much $a \times b$ differs from $b \times a$—is called "quantization." Dirac's idea was to introduce a corresponding procedure for the electromagnetic field, which is governed by Maxwell's equations.

A particularly useful form of the Maxwell equations can be obtained by a mathematical transformation known as Fourier analysis. The French revolutionary activist and mathematician Jean Baptiste Fourier had discovered the remarkable fact that functions of quite arbitrary form could be written as a sum of infinitely many sinusoidal wave trains whose wavelengths are smaller and smaller fractions of a fundamental wavelength (or their frequencies larger and larger multiples of a fundamental frequency). The physical manifestations of this mathematical fact are familiar from our enjoyment of music—every kind of sound can be represented as a superposition of sound waves of various frequencies. Similarly, every kind of light can be decomposed into its various colors, as Newton had discovered, and each of these individual components has a single, specific wavelength. If we subject a given solution of Maxwell's equations to such a Fourier analysis, we may ask, are the monochromatic constituents of that solution subject to simpler equations? Since a sinusoidal wave is the plot of the very simplest kind of oscillation there is—namely, that of an ordinary pendulum bob on a string, gently swinging back and forth—it will not surprise you that each individual component of a solution of Maxwell's equations satisfies the equation of motion of a pendulum, also called a *harmonic oscillator.* In other words, any electromagnetic wave may be mathematically regarded as a collection of entities each of which satisfies the Newtonian equation of motion of a harmonic oscillator.

In many cases, classical mechanics and the quantum theory give fundamentally different answers to the question, what energy can the given system have when it is in a steady, unchanging state? Whereas classically a system can have a continuous range of energies, the quantum equations of motion will, in many circumstances, permit only a discrete set of energies. For a given harmonic oscillator, the procedure of quantization is especially easy to carry out, and the set of allowed energies is very simple: there will be a specific minimal universal "zero point" energy, and all the others will differ by steps of equal size—they are whole-number multiples of one "quantum" added to the zero-point energy. As a consequence, the operators emerging from the equations, which raise the system's energy from one level to the next, may be viewed as mathematically formulating the *creation* of one quantum of energy, the numerical value of which is equal to the frequency of the oscillator multiplied by Planck's constant.

What I have described here, on the basis of the quantization of Maxwell's equations, turns out to be the QED explanation of Einstein's light quanta or photons! If we start with Maxwell's classical equations for electromagnetic waves and subject them to the mathematical procedure of quantization in their Fourier-analyzed form, we are led to the conclusion that light of a given color or frequency always comes in the energy packets postulated by Planck and Einstein. It is noteworthy that this simple interpretation of monochromatic light consisting of independent photons, each with the same energy, is ultimately the mathematical result of an "accidental" feature of the quantum-mechanical equation of a harmonic oscillator—namely, that its allowed energy levels are all equally spaced, a feature of no other system. The energy levels of an electron in the hydrogen atom are not at all similar; we cannot picture such an atom in its fifth excited state after a collision as simply consisting of a bag of five independent quanta of energy added to the atom in its ground state. However, our result also contains a very disagreeable feature: since a general solution of Maxwell's equations requires the summing up of infinitely many Fourier components of different frequencies, it has to be thought of as a collection of infinitely many oscillators, and since each oscillator has the same zero-point energy, the sum of these energies is always infinite! This is one of the "divergencies" that beset QED and that had to be overcome by some rational procedure before the theory could become acceptable.

So far, I have discussed only the quantum theory of free electromagnetic waves, without paying any attention to the sources of these waves or quanta. According to the Maxwell equations, the sources of the waves are electric charges, such as electrons. When not emitting radiation, these electrons are, in turn, supposed to be governed by the Schrödinger equation or, if relativity and spin are taken into account, by the Dirac equation. Making the combination of electrons and associated electromagnetic fields consistent required another step. Like the Schrödinger equation, the Dirac equation for an electron in a hydrogen atom is a differential equation for the wave function of the electron. This equation was now subjected to a "second quantization," a procedure that replaces the numerical wave function by an operator, just as the numerical electromagnetic field function in the Maxwell equations had been replaced by an operator. As a result, we now have a quantized *matter field* in addition to a quantized electromagnetic field. The interaction of the two is expressed by the fact that the electromagnetic field enters into the Dirac equation (as a force) and the matter field enters into the Maxwell equations (as a source). That total combination of resulting nonlinear equations for the two fields constitutes QED, a quantum theory of fields that, after many mathematical difficulties such as "divergencies" were resolved by Feynman, Schwinger, and Tomonaga, yielded astonishingly precise numerical agreements with experimental results. The matter field in this theory, when considered in isolation, and when all coupling (the electric charge of the electron) to the electromagnetic field is imagined to be turned off, can be viewed as an operator that mathematically "creates" an isolated electron, just as the electromagnetic quantum field "creates" a photon. In the absence of these couplings, the electrons, like the photons, are all independent and know nothing of the presence of the others. With the couplings "turned on," however, all these particles interact with one another—electrons radiate photons, photons push electrons around, and photons scatter each other (an extremely small effect called Delbrück scattering of light by light), as do electrons—and the total conserved energy of a set of n electrons is not just the sum of their n individual energies in isolation but necessarily includes the energy of the photons.

We now know of a number of other quantum fields, which arise from various inter-particle forces discovered during the last sixty years; however, since these forces are much stronger than the electromagnetic ones—the coupling constants are much larger—calculations based on the assumed field equations are extremely difficult and not yet able to give results that can be confirmed by experiments with any accuracy. In addition, it should also be mentioned that, the remarkable success of QED nothwithstanding, there are still great mathematical difficulties with the kind of *local* quantum field theory I have described, in which the field operator (in the original, non-Fourier-analyzed form) can be thought of as the mathematical symbol for "creating" a particle at a point in space; the point may have to be replaced by a small region. Nevertheless, let us look at what kind of conclusions we can draw from this schema, which physicists now regard as the most basic of all theories, for a description of "reality."

My View of Basic Reality

The foregoing discussion, and that of the last chapter, should have convinced you that neither particles nor waves are appropriate entities to populate the ontology of the physical world at the fundamental level. They are practically unavoidable for conceptualizing everyday reality, including objects as small as those visible only through a miscroscope, but the quantum phenomena have taught us that the language and the notions suitable for physical descriptions at that scale are inadequate for the purpose of acquiring a more basic understanding of matter. The view I present, therefore, is one in which the most appropriate tool for the description of reality on the sub-micro scale is the quantum field, to be regarded as a "condition of physical space." In contrast to the wave function, the quantum field does not "live" in configuration space, whose dimension depends on the number of particles, as you may recall, but in three-dimensional space. This condition of space is more complicated than that of a classical field, which was already complicated and abstract enough; there is, however, no difference *in principle* between a condition of space that can be represented as a set of numbers and one that requires a mathematical description as an operator.[16] Indeed, there are ongoing attempts at "quantizing" the very notion of geometry itself, which may ultimately lead to a solution of the deepest open problem in physics, the reconciliation of gravity with the quantum theory. The sum total of all the field operators "at a point in physical space and time" (subject to the *caveat* mentioned in the last paragraph) comes as close to representing the fundamental reality as we can get; neither particles nor waves are in evidence because the theory requires none at this level: it is a thoroughgoing field view. "The two entirely unrelated classical conceptions of discreteness of particles, of continuity of the field," as Schwinger describes the quantum field, "now are unified in this entirely new conception; if not unified, then transcended."[17] Moreover, the fields develop deterministically: the field equations are partial differential equations of the first order in time, which means that, like Hamilton's equations of motion for particles in classical mechanics, they determine the fields in the future if they are known now; and relativistic causality is taken into account in the guise explained in Chapter 8.

In order to make contact with this reality, however, we have to observe it, and the reports of our observations must necessarily be couched in terms of everyday or classically describable entities like waves and localizable particles. All statements about the consequences of the theory—including probabilities of observing particles, their number, or their momenta—can be readily calculated from the fundamental fields by well-established rules of the quantum theory. In this picture, the particle-wave duality entirely disappears from the basic physical description of what is real; particles and waves, and the puzzles associated with them, enter only in the description of observations or measurements, for which we have no choice but to use a language suitable for the apprehensions of our senses.

In 1958, Heisenberg, replying to Schrödinger, made the following somewhat cryptic remark:

> Only the waves in configuration space . . . are probability waves in the usual interpretation, while the three-dimensional matter waves or radiation waves are not. The latter have just as much and just as little "reality" as the particles; they have no direct connection with probability waves but have continuous density of energy and momentum, like an electromagnetic field in Maxwell's theory.[18]

It is not clear from the context exactly what he had in mind when referring to "three-dimensional matter waves," but if he meant the quantum field, as seems likely, I am advocating a similar view. My argument concerning the inadequacy of our language to describe reality also resembles his earlier call to give up "Anschaulichkeit"[19] (visualizability) at the submicroscopic level. However, Heisenberg placed the particle concept at a more basic level; for him, particles always came first. In contrast, Schwinger asked: "Is the purpose of theoretical physics to be no more than a cataloguing of all the things that can happen when particles interact with each other and separate? Or is it to be an understanding at a deeper level in which there are things that are not directly observable as the underlying fields are, but in terms of which we shall have a more fundamental understanding?"[20] In my view, particles are no more than manifestations of the description of our observations. While they are not exactly our own creations—in some sense they are out there—they are not part of basic reality. Particles themselves may be regarded as secondary qualities of the field, analogous to the properties John Locke assigned to them. Their existence does not depend directly on our bodily senses—sight, touch, smell, taste, and hearing—but it does depend, in part, on extensions of these senses by the most sophisticated and powerful measuring devices. Does this mean that the Moon is not there when I am not looking, or the desk on which I do my work disappears when I am out of the room? Not at all; reality at the everyday level differs from that at the sub-micro level, connected though the two are. The complicated paths leading from the field to "particles" in the quantum sense, and from these "particles" without identifiable trajectories to bulk matter that we can touch, carry, mold, and use to build houses, are in principle well defined.

Reality at the submicroscopic level, then, so far as we can ascertain it, consists entirely of quantum fields, and all the wave-particle paradoxes are consequences of our need to use an everyday language to describe a reality for which this language is inadequate. The "weirdness of the quantum world" does not disappear, of course, but its origin should be sought in the mismatch between basic reality and our linguistic equipment, rather than in that reality itself. Recall the words of Niels Bohr: "There is no quantum world. There is only an abstract quantum mechanical description." If "quantum world" is meant to refer to a world populated by particles and waves, I must agree; if it refers to the deeper world of the quantum field, I cannot go along, for in that world there would be no reason why our mathematical description, in the unintuitive language of quantum fields, should not reach reality. A quantum field may appear very abstract for what is "out there," but that is the way Nature is, and we cannot change it. Human language and intuition have evolved from experiences accessible to our sensory facilities at the macroscopic level; it would surely be surprising if they were suited to those parts of Nature that our senses cannot reach.

There is no suggestion here that the quantum field is in any philosophical sense "ultimate reality"; Kant's *Ding an sich* can never be contacted and, hence, is scientifically meaningless. The reality of the quantum field is *unanschaulich,* and any other mathematical description that may take its place in the future will be as well. But submicroscopic reality is not forever closed to us just because our intuitive conceptual instruments are inadequate; the language and tools of mathematics are much more powerful, for they allow us to probe beyond the reach of everyday-scale notions. This is, ultimately, why mathematics is indispensable for physics.

11

TRUTH AND OBJECTIVITY

IN this final chapter I return to the chief objective of this book, which is the discussion of *truth*. Let me first briefly recall and summarize the background furnished by the earlier chapters. The primary tools of the physical sciences, whose central aim is to understand and explain natural phenomena, are theories and laws leading to successful predictions subject to observational or experimental test. Facts form the basis on which these theories ultimately rest, but there are many occasions on which facts and theories are intertwined. Originating in the human imagination, theories are subject to a variety of esthetic, psychological, social, and cultural influences, which are important from a historical point of view. But these influences are ultimately irrelevant for their acceptance, endurance, or rejection. Falsifiability and successful verification by publicly accessible evidence determine which theories flourish and which die.

The school of conventionalism asserts that at least parts of viable scientific theories are conventions, and indeed I have shown that to be so in some cases. But this view of science does not stop there; a particular, recent variant of conventionalism contends that scientific theories and findings are specifically the result of social and political influences and pressures. Some of its most prominent adherents go so far as to claim that *all* scientific statements, and even the facts themselves, are social constructions—the totality of which we call "Nature"—and that these constructions are unrelated to anything in external reality. There can, by now, be no doubt about my strong disagreement.

Of the explanatory principles of science, the most potent is that of causality. But whereas the Aristotelian notion of efficient causes was effectively demolished by David Hume, its successor, in the form of constant or statistical correlations—including a delay between the occurrences of a cause and its effect—continues to play a crucial role. The quantum theory introduced, for the first time, the concept of probability into science at a fundamental level, simultaneously destroying causality for the entities—waves and particles—in terms of which classical physics had been formulated and denying that the processes involving them could be described causally without the introduction of strange, instantaneous, long-range influences. While this theory gives all waves a corpuscular aspect, combining it with the theory of relativity makes the notion of fundamental "particles" itself both nebulous and puzzling. I concluded that the underlying reality had to be described in terms of the quantum field, and that the weirdness of the quantum world is the result of an incongruity between this reality and the language in which observations have to be expressed. We are, then, left with the most basic question: is science as described here *true*, and what, exactly, does this mean?

Different Truths in Different Disciplines

To begin with, we all recognize there are various kinds of truth, but this emphatically does *not* mean that "all truth is relative." What a religious Catholic has in mind by asserting the truth of a pronouncement by the pope, what a reader means by appreciating the truth of Dickens's *Bleak House* or a sonnet by Shakespeare, what an onlooker means by finding truth in Picasso's painting *Guernica,* and what a scientist means by being confident of the truth of the BCS theory of superconductivity—all these are different things. Without wishing to discountenance the value of different truths in their own sphere, the only kind of truth I will be concerned with is that of science. The words used for "truth" in the scientific, religious, spiritual, historical, esthetic, or artistic sense happen, unfortunately, to be the same, but they are incommensurable; no one has the right to deny claims that a statement is true in one meaning of the word because it is false or lacks meaning in another.

For a more concrete example, take the assertion by some opponents of the theory of evolution that their postulation of the existence of a Creator to explain the variety of life on Earth is a rival scientific theory to Darwinian evolution. As pointed out in Chapter 4, the theory of evolution, in some form, is a falsifiable scientific theory with a large body of evidence in its favor. One may deny the force of this evidence and doubt the theory without leaving the realm of science, but the meaning of the claim, made by some critics, that there exists a unique Being ultimately responsible for life is of a different character altogether. If this Being is part of the physical universe, we can legitimately ask scientific questions about her properties, the efficacy of her interaction with the rest of the universe, or her history. The scientific question of the origin and development of life is therefore not answered by postulating a Creator but is simply transferred to another level; life is explained by a force that itself requires explanation. Such questions, however, are rejected by these opponents of Darwinism. The unique Creator, by virtue of His divinity, can have no truck with the tests and criteria of science, which means He is scientifically *ad hoc* or redundant. Thus "creation science," whose truth is meant to be of another order, is not, after all, acceptable as science; its aim is unscientific and the two varieties of truth are not commensurable.

On the other hand, what of those who see a divine Creator at work before the Big Bang? If that initial instance was indeed a *singularity* in the real sense, science can have nothing to say about any "before"; there is no scientific reason to oppose arguments whose validity rests on other truth criteria, even if these arguments conflict with some cosmologists' speculations that are themselves of dubious scientific merit and have primarily emotional value.

Goethe's opposition to Newton's theory of light and color represents another example of the collision of truths emerging from different realms. The great man of literature, near the end of his life, regarded his "scientific" work of more lasting importance than his poetry. He disagreed with Newton's theory that white light consisted of a mixture of colored components on the grounds that it represented a dissection, violating Nature's wholeness and the poet's profoundest esthetic values. More than a century later, in Nazi Germany during the Second World War, this clash of the venerated giant of German Romanticism with Britain's brilliant genius of cold reason had great emotional and political ramifications. Heisenberg tried to assuage the conflict in a public lecture in Budapest in 1941, by calling into question the existence of a real contradiction: Newton was right within science, he said, while Goethe's truth was of a different order. To agree with Newton as far as physics was concerned did not oblige you to deny Goethe's approach to Nature, which was based on experience of another kind. Though he surely violated Goethe's intention, Heisenberg thereby skillfully exploited the different meanings of truth in science and humanism in order to avoid a conflict damaging to both.

What is Truth in Science?

When we speak of the truth of something, the first point to note is that this something has to be a statement or an *assertion;* contrary to frequent usage, it makes no sense to speak of the truth of a fact or of a property.[1] "The 'facts' themselves . . . are not *true.* They simply *are,*"[2] William James reminds us. To insist on this is not pedantry or hairsplitting. Formulating an assertion is attempting to communicate and therefore requires transmissible concepts and language: truth thus cannot be separated from human concepts and our linguistic apparatus. The history of philosophy is permeated with controversies that are of purely linguistic origin; apparently "deep" statements made in the language of one culture sometimes are untranslatable into that of another for want of a grammatical construction.[3] Awareness of the pitfalls of language should make us cautious when dealing with "truth."

Furthermore, it is useful to distinguish between the *definition* or the *meaning* of truth on the one hand and *criteria* for discerning it on the other. To be given a roadmap for finding the Grand Canyon is not the same thing as being told what it *is.* You may ask, Unless I know what it *is,* how will I recognize it when I get there? But science will not provide a shiny plaque, engraved with an eternal description of *The Truth;* it will furnish only a powerful torch to help in the search and a sense of smell to tell when you get closer. Since this book is not a treatise of philosophy, my primary concern will be with criteria, not with an abstract definition of truth, though it will help us also to look briefly at the latter.

What we most commonly mean when we claim an assertion is true is what Aristotle meant: it "corresponds to the state of affairs" described in it; a true statement is a "representation of nature," or a mirror of it. The statement "this flower is red" is true if and only if, in fact, this flower is red. The logician Alfred Tarski[4] even incorportated the "correspondence theory of truth" into formal logic. While to non-philosophers such a definition appears so obvious that it hardly needs mentioning, its meaning becomes more obscure in the context of fundamental science. The "state of affairs" described by "this flower is red" is clear enough, but it becomes less so when the statement is a scientific theory, especially when we recall that most theories are abstractions and simplifications with only approximate applicability in Nature. Is the "state of affairs" described by Newton's laws of motion a clear concept? What about the state of affairs in the EPR experiment? Although there is surely something right about the correspondence theory, as we ponder its significance it grows dim and murky. Because defining truth is not my main interest here, I won't dwell on its meaning further, especially since correspondence is difficult to use as a criterion.

One of my primary concerns is to ask whether truth is *relative:* are there, within science, varieties of truth—is truth in the eye of the beholder? An affirmative answer to this question lies at the core of the "perspectivist" theory of science and all the other relativistic[5] opinions described in Chapter 2. The theory of truth at the basis of the constructivist approach is that of *consensus.* In this view, consensus among the members of a culture is not only a good way to get to the truth—within limits, a defensible proposition—but *defines its meaning;* in other words, consensus is its only criterion of validity. This is why constructivists, who consider science but an edifice erected by no other means than agreement among an in-group, regard Nature "out there" as irrelevant to the final determination of scientific truth. Indeed, according to them, Nature *is* nothing but this construction. Recall the proclamation by Latour and Woolgar: "Scientific activity is not 'about nature', it is a fierce fight to *construct* reality."[6] From this viewpoint, science is like a scaffold put up ostensibly for the close study of prehistoric paintings on the walls of a large cave; when you manage to penetrate the dense grid of girders and ladders, you discover that there are no paintings—the scaffold has been erected for its own sake and the artwork consists of no more than reports from the experts studying it.

Such a definition of truth leads inevitably to relativism, since the inhabitants of various eras or cultures are likely to adopt different truths and there is no way to compare or distinguish them transculturally. "We are prisoners caught in the framework of our theories; our expectations; our past experiences; our language," Karl Popper grants.

> But we are prisoners in a Pickwickian sense: if we try we can break out of our framework at any time. Admittedly, we shall find ourselves again in a framework, but it will be a better and roomier one; and we can at any moment break out of it again . . .
>
> The Myth of the Framework is, in our time, the central bulwark of irrationalism. My counter-thesis is that it simply exaggerates a difficulty into an impossibility. The difficulty of discussion between people brought up in different frameworks is to be admitted. But nothing is more fruitful than such a discussion, than the culture clash which has stimulated some of the greatest intellectual revolutions.[7]

For the philosopher Richard Rorty,[8] the search for truth is nothing but "edifying conversation"; he ignores the fact, emphasized by Ian Hacking,[9] that it forms a basis of action. There is little sense in pretending that truth has no consequences; it has different kinds of consequences in different realms. In science, "the influence of knowledge over action arises from its power of prediction,"[10] John Ziman rightly points out, and this influence is considerable. "Conversation," or story telling, is hardly an appropriate way to describe our search for it.

There is another reason why the consensus theory of truth is inadequate. I grant that the assertions of science, accepted into its body of truths after much testing, eventually become consensual, and that they must therefore be, in Ziman's terminology, *consensible*—formulated so as to be capable of being accepted or rejected on the basis of a scientific consensus. This is rarely their final status, however; they are always subject to a residue of potential doubt. If agreement were the *definition* of truth, consensible scientific statements would not be as precise as they usually are but would tend to be as vague as possible. We see examples of the value of vagueness every day in politics and international diplomacy—the less precise, the more likely an agreement. Furthermore, while general accord might do justice to the internal usefulness of truth among those consenting, it cannot account for the great functional value of scientific truth in the external world.

From the viewpoint of psychology, the consensus theory may be regarded as a manifestation of behaviorism grown rampant.[11] Just as behaviorists define psychological states, feelings, and thoughts entirely by their outward signs in behavior, so the truth is defined by its behavioral demonstration—communal agreement. As a definition or as a sufficient criterion of truth, the consensus theory has to be rejected as a chimera.

The proverbial attempt by three blind men to describe an elephant may serve as a metaphor for our purpose: one man touches the beast's trunk, the second its tail, and the third a leg. The resulting descriptions of the animal will, of course, be incongruous and difficult to reconcile. Truth has many aspects, and even within science it may be approached from distinct angles with varying outcomes. Recall, for example, the two apparently different versions of quantum mechanics developed almost simultaneously by Heisenberg and Schrödinger. In Chapter 6, I rejected the inductive model of scientific theories—a clear-cut, unique process of induction leading from the data to a theory—advocated by philosophers of science for many years but no longer in fashion. More congenial was Popper's deductive view that theories are products of the human imagination, their validity depending on agreement between their consequences and observational facts. We therefore have no guaranty that there might not be several rival theories of equal validity. Are they all equally true? Would we not expect the scientific theories of Barrow's aliens from Alpha-Centauri to differ from ours? Does that leave the truth in the eye of the beholder?

In answer, let me begin with the observation, agreed upon by most scientists, that a theory is never true in an ultimate sense: it can, at best, only approach the truth. Thomas Kuhn[12] initiated the contemporary vogue of relativism among sociologists of science by drawing on the analogy of Darwin, whose most revolutionary concept was the total absence of a goal in biological evolution. Similarly, Kuhn argued, the progress of science does not move *toward* anything such as truth; it simply evolves. This implies that we should expect the science of Barrow's aliens to be incommensurate with ours, which would be hard to reconcile with their ability to build space ships or long-distance communications, as we do. The science philosopher, Abner Shimony, replied to Kuhn's argument with an analogy (going back to Descartes) between attempts to discover the laws of Nature and attempts to decipher a message encoded in a text.

> Suppose that we have such a text, and that after numerous conjectures the tentative decipherment has become more and more coherent. The success may be no more than a series of coincidences, so that the tentative decipherment is completely on the wrong track. But somehow it is more plausible that a good approximation to the correct rules of encoding has been found than that the long run of successes has been coincidental. Kuhn's thesis that the truth plays no role in the progress of science is analogous to maintaining that progressive coherent decipherment could occur even though no such things as the initial message and the rules of encoding exist.[13]

The Criterion of Coherence

In referring to the metaphor of the three blind men, I ignored one crucial property of the elephant. Though the three observers perceive the nature of the animal differently because each touches a part of the same creature, they are not permanently rooted to their positions but can grope their way along the skin toward the others, coming at last to the conclusion that what they thought were different entities, or incommensurable versions of a single entity, made up one coherent whole, an entire elephant. This is the central point for the recognition of truth.

Jacob Bronowski recounts a story from Eric Shipton's description of his ascent of Mount Everest in 1953 with his Sherpa, Angtarkay, as he climbed the mountain, familiar to him from the north, and saw it from the south for the first time: "I recognized immediately the peaks and saddles so familiar to us from the Rongbuk (the north) side . . . It is curious that Angtarkay, who knew these features as well as I did from the other side and had spent many years of his boyhood grazing yaks in this valley, had never recognized them as the same; nor did he do so now until I pointed them out to him."[14] It was indeed a single mountain that revealed diverse characteristics when viewed from different valleys. The recognition of this fact is essential for arriving at a coherent image of the region; without it, we have nothing but a jumble of inchoate impressions.

The most important criterion for ascertaining the truth of a statement is its *coherence* with a network of assertions that are also regarded as true. Clearly, this cannot serve as a definition of the truth of a proposition in isolation, nor as a method for determining whether an isolated report is true; it can serve only as a criterion for recognizing the truth of a whole body of statements, and of individual assertions within a larger context. William James understood that coherence was an important part of the pragmatic criterion—"consistency between the present idea and the entire rest of our mental equipment, including . . . our whole stock of previously acquired truths"[15] gives us satisfaction; it leads to *order.* And what is truth in science but "an ordering of the facts," Bronowski suggests. "We organize our experience in patterns, which, formalized, make the network of scientific laws." But neither science nor we live

> by following a schedule of laws. We condense the laws around concepts. So science takes its coherence, its intellectual and imaginative strength together, from the concepts at which its laws cross, like knots in a mesh. Gravitation, mass and energy, evolution, enzymes, the gene and the unconscious—these are the bold creations of science, the strong invisible skeleton on which it articulates the movements of the world.[16]

Scientists use the test of coherence in a great variety of circumstances, not only in a technical context—judging whether a new local theory coheres with another of greater generality—but often in more general contexts, as a result of which opponents with unorthodox ideas sometimes complain their views are not being taken seriously. Supporters of parapsychology, for example, have a long-standing grievance that scientists refuse to pay serious attention to what they claim are results of their "scientific" experiments. Researchers justifiably refuse to listen to these claims, to examine or refute them in detail, because they are incoherent with the rest of our scientific knowledge, but in doing so they open up the scientific community to charges of "elitism" and closed-mindedness. It is possible, of course, that a startling, discordant fact, long ignored by the "establishment," may someday be discovered, producing a new paradigm with its own coherence. The body of scientific knowledge, however, is by now so large that this scenario is extremely unlikely, and examining seriously and carefully every assertion that falls far outside the coherent web of scientific learning would hardly be productive. An entirely new effect would require an enormously convincing demonstration to persuade scientists to pay attention, but if the effect is real, it will eventually be embraced. Overcoming the natural conservatism of scientists is difficult but not impossible.

The criterion of coherence of the body of scientific truths rests largely on the cliché that science *works.* This notion comprises many disparate ideas: (1) the concepts and explanations of science have practical consequences that may be observed by everybody—television and computers are outgrowths of technology we all use every day; (2) the predictions based on scientific theories come true—later observation confirmed the prediction of the general theory of relativity that starlight would be bent by the sun, and the prediction of the quantum theory that a beam of electrons should be subject to diffraction was verified in fine detail by experiments; (3) when a well-corroborated theory implies that a certain phenomenon will never occur, it will, indeed, not happen—the laws of physics foreclosed the possibility of engines that run without input of energy, and, in spite of many attempts, none has ever been constructed. Contrary to the sociological interpretation, the phrase "science works" is not intended to mean that it works toward some social end or to someone's advantage; the question asked by Evelyn Fox Keller—"After all, for whom or for what can a science that provides tools powerful enough to destroy even their own makers be said to work?"[17]—is quite beside the point. "Science works" most of all *intellectually.* That is, it provides a rationally coherent structure of comprehending the world, and this comprehension has far-reaching practical effects, including its coherence with the external world of our experience and the ability to make successful predictions on a regular basis.

Even though parts of the edifice may be found to be rotten, the coherence of a body of scientific truths accounts for its stability over long periods of time. "Scientific knowledge," John Ziman rightly observes, "eventually becomes a *web* or *network* of laws, models, theoretical principles, formulae, hypotheses, interpretations, etc., which are so closely woven together that the whole assembly is much stronger than any single element."[18] In contrast to many other kinds of truth, the truth of science does not rest on permanent acceptance of any one of its parts. It has no scripture whose every word must be believed to preclude the collapse of the whole, nor do scientists expect even their most cherished theories to remain valid forever. The *structure* of science, however, will endure.

For scientists, the edifice of truths they erect is beautiful, and such intellectual beauty, Poincaré argues, motivates them in their "long and difficult labors," more than "the future good of humanity."[19] It is precisely the admirable coherence of the architecture scientists see before them that many find sublime. Others stress the *simplicity* of the truth of science, but they are merely voicing their preference for a coherence in which there are no extraneous elements and *ad hoc* curlicues. In other words, both beauty and simplicity are attributes of the coherence of scientific truth.

Using coherence as the primary criterion for truth raises the question of whether several competing bodies of truth might possibly coexist, each consistent within itself. It is undeniable that internally coherent systems of one kind or another have appeared in the past. Indeed, there are those who make much of the internal coherence of belief systems based on myths or folklore and suggest they are equally valid competitors of modern science. The meaning of coherence in such arguments, however, is always restricted and never extends to the entire external world of experience. Astrology does not "work," except in the imagination of its adherents, and neither does witchcraft. Early versions of science were internally consistent bodies of purported knowledge, but they were superseded because they lost their consistency with the external world when new facts were found. No doubt, our science will lose some of its present coherence when future scientists make their discoveries. That is why today's science is closer to the truth than medieval science was, and the science of the twenty-second century will be closer yet. It is not that Babylonian science "was true then and is false now," that our science "is true now but will be false a thousand years from now," but that we are approaching the truth ever more closely, Thomas Kuhn notwithstanding. Science *does* make progress.

Reconciling Truth and Imagination

How can a gradual, progressive approach to truth be reconciled with my earlier assertion that scientific theories are products of the human imagination that simplify Nature in ways that are not uniquely determined by the "facts"? How does it reckon with the recognition that there are conventional elements in the laws of Nature? Barrow's aliens are not likely to have exact analogues of Newton's laws of motion or of our quantum mechanics. It is, above all, because *no theory fits Nature precisely that there is room for a variety of theories to account for a large body of facts.* Some physicists[20] argue that we have a good example of two such competing theories, each at variance with the other and of equal range: the orthodox version of quantum mechanics and the quite different theory, promulgated by David Bohm and his followers, that contains unobservable "hidden variables" acting causally but non-locally. Bohm's theory is ignored by most physicists because the fact that its predictions seem to agree with those of conventional quantum mechanics makes it pointless to them. Its view of reality, on the other hand, while no less puzzling because of its non-local character, differs greatly from the Copenhagen interpretation. Discrepancies of this sort are one reason contact with an alien civilization would be of enormous interest. If a scientific theory were true in any ultimate sense—corresponding to "the state of affairs" exactly—there could be no interpretive disagreements, but few scientists believe theories represent rock-bottom truths.

The quantum theory teaches us that our conceptual and linguistic apparatus, based as it necessarily is on everyday-scale experience, is poorly matched to the reality far below that scale, and I have argued that the mismatch results in a weird and acausal appearance of the submicroscopic world. It might therefore be thought that if Barrow's aliens were billions of times smaller than we are, their experiences would be quite different from ours and their concepts would be better descriptions of the truth about reality at the submicroscopic scale. A serious science-fiction writer presenting a world inhabited by such tiny intelligent creatures, however, would betray his ignorance of physics. It is no accident that evolution has produced thinking animals of our size; there are good physical reasons why we could be neither larger nor smaller by many factors of ten. Barrow's Alpha-Centaurians would necessarily have the same difficulties conceiving of submicroscopic reality we have. That truth and reality become puzzling and conceptually hazy at the sub-micro scale is intrinsic to the structure of the world and to the evolution of life conceivable anywhere. It is not a purely human failing.

What kind of scientific assertions, then, can be regarded as true? Keeping in mind that we are not searching for ultimate and eternal truths but always for truths that are provisional and approximate, we must acknowledge important distinctions to be made among assertions of facts, theories, models, metaphors, and analogies. Because they recognized that statements of individual facts, describing impressions of our senses and their extensions in the form of powerful instruments, are the safest truths, some positivists attempted to base science on nothing but records of "pointer readings." As we know, however, *individual* facts are of relatively little interest to science; statements of *general* facts are the ones that count—"the mass of the proton is 1.67×10^{-24} grams," "the speed of light is 3×10^{10} cm/sec," "matter is made up of atoms and molecules," etc. Because of their universality, the correspondence definition of truth does not work well with such assertions. Their intuitive and heuristic significance is captured most fully not by giving them an instrumental meaning—defining them in terms of the experimental or observational procedures used for their verification—but by relying upon their coherence within a large body of other factual statements and theories. We believe in the truth of the atomic theory of matter not because of any one specific experimental result, nor could any individual experiment dissuade us from our belief; we believe in its truth because it coheres with an enormous set of different observations in various parts of physics and chemistry. To deny it would seem utterly quixotic, and we find it hard even to credit the doubts expressed by a respectable physicist like Ernst Mach as late as the beginning of this century.

It is clear that models, metaphors, and analogies lack the attribute of truth. As noted earlier, the word *model* is, in fact, usually employed to emphasize this deficiency; metaphors and analogies can at best be regarded as suggestive or as true in a non-scientific sense. But what about the truth of the theory of relativity, or of Maxwell's theory of electromagnatism? What of such statements as "light consists of photons" and "matter is made up of waves"? As I noted earlier, theories are usually imaginative mental constructions, corroborated by agreement between their implied predictions and observational facts. It is very difficult to see them corresponding to a "state of affairs" in the external world with which they deal. Since they are always simplifications and abstractions of the relations between facts and events—incorporating even elements that are conventions—to what extent can they be regarded as *true?* These amalgams of simplifications, abstractions, and conventions are simply convenient. "But it is true that [they are] convenient," writes Poincaré, "it is true that [they are] so not only for me, but for all men; it is true that [they] will remain convenient for our descendants; it is true finally that this cannot be by chance."[21] In other words, the very fact that certain conventions and simplifications are convenient and useful says *something* about Nature and reality; it is therefore hard to deny them the status of truth, at least in a tentative and partial sense. Like ours, the theories of the Alpha-Centaurians capture a part of the truth, just as each of the three blind men has a hold on part of the elephant.

When it comes to formulations of results of the quantum theory in terms of waves and particles, however, we are caught in the linguistic quandary alluded to earlier. Since truth is a property of statements necessarily made in a language evolved from experience at the everyday level, inappropriate for the submicroscopic world, we cannot call such assertions of the quantum theory true. It is impossible to find an unambiguous "state of affairs" to which they correspond, and what is more, the lack of complete coherence constitutes a notorious problem with the quantum theory. Since the description of matter in terms of particles such as atoms, molecules, neutrons, protons, electrons, and so on exhibits coherence over a very large area of phenomena, it would be folly to withhold from it the attribute of a partial truth. However, neither of the statements "electrons are particles" or "electrons are waves" can be said to be true; on the other hand, the statement "electrons manifest themselves sometimes as particles and sometimes as waves, depending upon the mode of observation we apply" is true, but it is not about reality but about how we perceive it. The only form in which it makes sense to assign the quality of truth to the description of the sub-micro world is that in which, as I argued earlier, this world has reality—the abstract and unintuitive assertions about the quantum field made in mathematical language. This does not imply that the quantum theory of fields can be expected to be eternally true and might not be superseded by another theory; as I have emphasized before, all scientific truth is provisional. But we cannot avoid the uncomfortable conclusion that some scientific truths about Nature require for their formulation an abstract language far removed from everyday verbalization; any attempt to translate them into more intuitive form makes them either false or meaningless.

There is an important lesson contained in this conclusion: although every truth requires a language in which to state it, this language does not necessarily employ the words and concepts of ordinary discourse. I don't mean to give aid and comfort to obscurantists—of which there are more than enough outside science—but at the level of the most basic physics we see an example of how impossible it is to state the truth of reality in intuitive language; every attempt to do so, indeed, ends up in paradox and confusion. The only vocabulary capable of serving the purpose of describing this truth appears to be that of abstract mathematics. "Whereof we cannot speak, thereof me must be silent," Wittgenstein announced, but unless he intended to include mathematical symbols in the category of speech, he was wrong. The very abstractness of mathematical language, divorced from *Anschaulichkeit,* the visible world, as it largely is, serves the indispensable purpose of communicating "whereof we cannot speak" in any other language. With mathematics we may approach truths about aspects of reality that would otherwise remain hidden. The notorious inarticulateness of many scientists when they try to communicate their ideas to the general public is therefore not entirely due to an inability or unwillingness to make themselves generally understood; it is, at least in part, intrinsic to their subject matter.

The issue of whether human thought processes require the use of language is a contentious one, which I can hardly resolve here. Percy Bridgman insisted that he did not use words in most of his thinking, and, since he opposed excessive abstraction, it is very unlikely that he thought in mathematical terms. But, whatever form scientists' internal ratiocinations take, they must eventually communicate to others the truth they believe they have found; private science is an oxymoron. To that extent the sociologists certainly are correct: science is, indeed, a social activity. Scientific truth, therefore, cannot be divorced from language in its most general meaning, including that of mathematics.[22]

Objectivity

The concept of objectivity is a signpost on the way to the truth. As I have argued earlier, the overriding characteristic of modern science is its reliance on openly ascertainable knowledge; its truth is, above all, *public.* Plato thought that the way to seek the truth is to consult a wise man, but the validity of scientific statements does not rest on the pronouncements of any guru or expert, though it may sometimes appear that way to outsiders. It rests on evidence that could, in principle, be checked by anyone with the needed fundamental knowledge and apparatus (which admittedly presents a hurdle difficult to overcome). For this reason, the Royal Society chose as its motto *Nullius in Verba,* an abridgment of Horace's *Nullius addictus iurare in verba magistri*—"not bound to swear by the word of a master."

A clinical psychologist once recounted her personal experience of seeing Elvis Presley in her office, alone, long after his death. "And by the time it was over," her account ends, "I understood that there is much more to the mind and the human spirit than I had previously allowed, and that if I was going to be a full human being and be helpful to others, I had to realize this and let it affect me fully. I instinctively bowed my head and put my hands together, as in prayer. When I looked up again, he was gone."[23] This is not how a scientist arrives at truth; private illumination is not its touchstone. Charles Sanders Peirce had it right:

> Unless truth be recognized as *public,*—as that of which *any* person would come to be convinced, if he carried his inquiry, his sincere search for immovable belief, far enough—then there will be nothing to prevent each one of us from adopting an utterly futile belief of his own which all the rest will disbelieve. Each one will set himself up as a little prophet; that is, a little 'crank,' a half-witted victim of his own narrowness.[24]

This open quality of the scientific approach to the truth is indispensable for objectivity, and *vice versa*—science must be open to general inspection, not held in secret ritual or private vision. Without public access, it would be impossible to assure objectivity or the absence of distortion produced by personal preferences and individual perspective. The unrestricted character of objectivity is quite the reverse of Bloor's definition of the word, quoted in Chapter 2, as "institutionalised belief." This definition surrounds the meaning of *objective* with the walls of a cloister, where a scientist might indeed become "a half-witted victim of his own narrowness," or the narrowness of the institution. To argue that social and institutional barriers cannot be overcome "simply exaggerates a difficulty into an impossibility"; we are not literally prisoners in a keyless dungeon. On the other hand, public access hardly means that everyone without the required background knowledge should be regarded as equally competent to judge the truth of scientific propositions; in that sense, and in that sense only, science is inevitably "elitist."

The ethos of objectivity that every proper scientist subscribes to requires that the search for truth about Nature be disinterested. If an experimenter is looking for observational facts, she has to accept what Nature offers, irrespective of whether it conveniently fulfills her expectations or awkwardly obstructs her pet hypothesis. If a theorist is constructing a conjecture and throws it open for testing in the laboratory, he finally has to assent to the verdict of those tests even if it destroys the theory he has spent years developing. He may resist the experimenters' results for a time, he may blame them on observational errors or other flaws so that he can continue to believe in the beauty and truth of his theory, but in the end he is obliged to submit to the judgment of Nature. Such demands on personal integrity are often hard to follow—Alan Chalmers is right in saying "*Objectivity is a practical achievement*"[25]—but these demands are ultimately the foundation on which science rests. The open nature of scientific truth cannot be maintained if there are systematic attempts by its practitioners to subvert it through the suppression of inconvenient evidence or the propagation of theories contradicted by the facts.

The very mode of expression used by scientists in their communications is influenced by their attempt to be, and to appear to be, objective. It is a mode of expression that often seems dry and colorless to outsiders. You are much more likely to find the bald statement "Matter is made up of particles" in a scientist's writings than "I strongly believe matter is made up of particles," but this has little to do with the author's strength of conviction. Any straight assertion of "X" opens X up for discussion and debate, whereas "I believe X" can only be countered by "I don't believe X." Unless violence ensues, saying "I don't believe X" would either be the end of the discussion or the beginning of an analysis to discover why someone like her *would* believe "X," and this in turn would be the beginning of the sociology of science. I do not mean to imply that there never is passion behind the statement of "X" by a scientist, or that the person stating "X" will not do his best to convince readers of its truth. Most scientists, in fact, feel strongly about the truth of at least parts of science, and discussions among them often become quite heated. In scientific discourse, however, the statement of "X" is never meant to be *ex cathedra*—even when the arrogance of some individuals may make it sound that way. It is simply intended to be divorced from the person making the statement. Keeping a distance between the statement and the individual making the statement also explains why scientists are notorious for writing up their results in the passive mode rather than in the more personal active voice: "X was found by Smith" distances the discoverer from the discovery more than "Smith found X" does.

The ideal of objectivity is today under heavy attack. Cynics point out that, contrary to the myths propagated to glamorize famous scientists, the ideal is violated in many concrete instances. They are right, of course—scientists are human, and since recognition of their achievements is their primary reward, they may "cut corners" to get credit for their work. Like others, they sometimes succumb to weaknesses such as jealousy, vanity, and, on very rare instances, even dishonesty.[26] Scientists are also famously competitive, and some feminists have connected this characteristic with the dominance of men rather than women in science. Competitiveness can, of course, sometimes undermine objectivity, but whether successful science is possible without it is an open question. It remains to be demonstrated that cooperative social models for science in place of competitive, individualistic practice would work as effectively. That parts of science have grown to involve large expenditures and others have rubbed against areas of application in which political passions run high has exacerbated such problems. Not all scientists have been able to remain aloof and objective, and many observers who habitually regarded them as superhuman have become disillusioned. But the failings of a few should have no bearing on the validity of the ideal and the need to maintain its pursuit as an instrument of enormous value to humanity.

To a large extent, however, science is notably stable with respect to the infusion of biases by its practitioners; that is the great virtue of its widely accessible character. "What is to be condemned," Jonathan Rauch rightly emphasizes, "is not bias but *unchecked* bias. The point of liberal science is not to be unprejudiced (which is impossible); the point is to recognize that your own bias might be wrong and to submit it to public checking by people who believe differently."[27] As he points out,

> the genius of liberal science lies not in doing away with dogma and prejudice; it lies in *channeling* dogma and prejudice—making them socially productive by pitting dogma against dogma and prejudice against prejudice.[28]

Science, in this respect, resembles capitalism, which also rests on steering destructive and undesirable human traits, like greed and avarice, into socially productive directions; both systems are remarkably fruitful. The pitting of "dogma against dogma and prejudice against prejudice" does not, however, diminish the importance of the ideal of objectivity for the enterprise as a whole. Nor would it justify rampant and egregious violations of disinterestedness by scientists on the grounds that the system is self-correcting. It is indispensable that objectivity remain as an aspiration for all contributors to the system.

If the cynics attack objectivity on the grounds of disillusionment, others do so on a political basis. "The scientific method rests on a particular definition of objectivity that we feminists must call into question,"[29] declares Ruth Hubbard, adding that "the pretense that science is objective, apolitical and value-neutral is profoundly political."[30] In fact, some critics of science go so far as to deny the very value and desirability of objectivity. According to these commentators, the ideal should not be to remain disinterested and objective but rather to be committed to general social betterment, justice, and whatever other purposes they deem worthy. To them, shouting "*I believe* in X" is much more important than "X." As Hubbard contends,

> Science and technology always operate in somebody's interest and serve someone or some group of people. To the extent that scientists are 'neutral' that merely means that they support the existing distribution of interests and power.[31]

Science is objective to the extent it avoids bias or external agendas, either because individual scientists are free of them or because the public character of science produces a balance with that effect. Those who question the objectivity of science, like Ruth Hubbard and others with political goals, claim that all assertions, including those of scientists, are necessarily colored by class, race, ethnicity, gender, religion, sexual preference, or any other suitable pigeonhole of the asserter. The goal of objectivity, in their view, is therefore unattainable and perhaps even undesirable, because knowledge is power, and this power will be used, if not for one purpose, then another. The great force of scientific knowledge is now applied to many objectionable and destructive ends; if scientists were guided by the right political consciousness, they would avoid seeking the kind of knowledge that could be used harmfully and would channel whatever knowledge they do obtain into socially desirable directions. Objectivity, which prevents scientists from following these paths, is therefore to be shunned, as these critics would have it.

That knowledge is power is indisputable, and so is the assertion that the fruits of science have been used for destructive purposes, as well as for many eminently constructive and beneficial ones. But there is no knowledge that is inherently sinful, and ignorance too can be exploited for evil political ends. To shut off and forbid certain questions for fear that the answers will be misused is to subvert the open character of science and to arrogate to social and political institutions a decision power based on ignorance. Such attempts have always failed in the past, and history has shown no good to have come of them, whether they were made by religious institutions or governments. When we censure scientists for producing potentially destructive developments that seem to them "technically sweet" or too interesting to set aside, no matter the consequences of their exploitation, we blame the messenger for the message and lock the door to all further deliveries. Remaining ignorant protects only the consciences, not the bodies, of those kept in the dark.

Values

The issue of objectivity has taken us into the contentious realm of values and their relation to science. To be sure, we must avoid the "naturalistic fallacy" of attempting to deduce ethical and moral prescriptions from the truths of science—that no *ought* can ever be implied by an *is* has been generally recognized ever since the writings of Hume. It is nevertheless clear that there are ethical imperatives implicit in the daily activities of scientists, some of which they surely carry with them when they leave their work. From a practical point of view, such implicit morals exert a more powerful influence on behavior than any explicit rules. Denigrated though it may be by those who esteem dedication to a cause more highly, objectivity is one of these values. There are certainly occasions when commitment to personal moral or political aims is worthy of great admiration, but no civilization could advance or even remain in existence for long if all its citizens led lives devoted to nothing but these fervent aims. Science and art would wither, and as Poincaré reminds us, "It is only through science and art that civilization is of value."[32] Indeed, without objectivity, even the pursuit of justice would be perverted into a search for vengeance. The argument that no one can be completely objective—we are all, to a certain extent, unavoidably biased, our thoughts and ideas tainted by our upbringing and social surroundings—has some foundation, but it is not sufficient to invalidate the effort to be so.

A concomitant of the scientific ideal of objectivity is the recognition that the source of genuine contributions is not necessarily restricted to any one culture, nationality, gender, or race, however much the contributions of these groups may have differed in the past. Of course, there have been many instances of individual prejudices among scientists and mathematicians. The mathematician Felix Klein thought that "a strong naive space intuition [was] an attribute of the Teutonic race, while the critical, purely logical sense is more developed in the Latin and Hebrew races";[33] the physicist Pierre Duhem depicted German scientists and mathematicians as lacking intuition[34] and characterized the English mind as "ample and weak" in contrast to the "narrow and strong" mind of the French.[35] Stark and Lenard championed "German physics" in opposition to the science pursued by the "Jewish mind" of Einstein, and the prominent Russian mathematician Lev Pontryagin is known for his extreme and active anti-Semitism. Such personal biases aside, however, natural scientists have almost always been in the forefront of efforts to fight the excesses of irrational chauvinism and bigotry. The creative work of scientists has been recognized across the boundaries of nations at war, hot or cold, and movements toward lessening international tension have often been launched by scientists. Soon after the end of the First World War, British scientists worked to come to a reconciliation with their German colleagues while Germany was still enveloped in a dark cloud of resentment and hate. International scientists began the Pugwash movement to ease the enmities of the cold war with the Soviet Union. It seems clear that these acts were largely the result of the scientists' strong attachment to rationality and objectivity. Though individual physicists have made significant contributions to the development of weapons, and the technology of war has often been based on scientific knowledge, nations have never been propelled to war for reasons of science, as they have been for religious causes.

The passionate search for truth has to be recognized as another value implicit in the scientific enterprise. I will, of course, be accused of naiveté for saying so, since we all know from many recently published biographies that scientists are rarely the pure lovers of truth depicted in movies and self-serving myths. In fact, they are often hatefully competitive with one another, and their search for truth is sometimes indistinguishable from a race for priority. Nevertheless, "The most vital factors in the method of modern science," Peirce writes perspicaciously,

> have not been the following of this or that logical prescription—although these have had their value too—but they have been the moral factors. First of these has been the genuine love of truth and conviction that nothing else could long endure.[36]

In contrast to most other human occupations and activities, science does not attract talented men and women with a promise of monetary gain, personal power, or public acclaim. The rewards for the pursuit of truth are the pleasure derived from our increased understanding of Nature and from the recognition by our peers. Political divisions and hatreds are minimized by the realization that the truth we seek is never final. "It is imperative in science to doubt," Richard Feynman tells us. "Every one of the concepts of science is on a scale graduated somewhere between, but at neither end of, absolute falsity or absolute truth."[37] No matter how strongly they believe they are on the right track while their opponents have lost their way, scientists will never launch a crusade to destroy the infidels under a banner of ultimate truth. Their flag is more likely to be inscribed with the words of the Enlightenment philosopher and poet Gotthold Ephraim Lessing:

> If God told me to choose between his right hand, which held all truth, and his left, which held the ever active search for truth, with the proviso that I would always and forever be in error, I would with humility kneel to his left . . .[38]

In the end, the relentless search for truth, never found, with the beacon of objectivity as the ideal, never quite achieved, engenders an *attitude* that pervades the lives of the men and women of science. To question authority is as natural to them as the air they breathe; *it is,* in a manner of speaking, the air they breathe. Furthermore, no other field of human endeavor is more dominated by what Bronowski calls the "principle of truthfulness."

The scientific attitude, which exerts its influence on society at large by the pervasiveness of modern technology, has been blamed by some commentators for many of the present ills of our culture. An excessive reliance on rationality is said to have destroyed the religious faith needed to sustain ethical and moral standards. In fairness, it is hard to deny that there is some justification for this assessment of the state of Western civilization. We see both a decrease of the influence of religion and a decay of moral values all around us, and no doubt the rise of science during the last four hundred years has been largely responsible for the decline of the power of religious faith in the West. Though they have contributed to the demystification of the universe, scientists have not been able to communicate very effectively to the public the awe many of them feel as their understanding of Nature grows. We can only hope that the present cynical and nihilistic state of our culture is a transitional one, that in the future it will be replaced by a more positive outlook, with the values of science in a central position.

In my judgment, however, it is a complete misreading of the history of this violent century to ascribe its ills to the ascendency of science, as some commentators do.[39] The greatest of the upheavals we have witnessed, the Second World War, was caused by a spasm of irrationality marking the end of the era of Romanticism. The Nazis did not rally under the slogan of "reason and truth" but *"blood and soil."* Theirs was a reaction against rationality, and most of this century's violence has been, and still is, driven by the myth-based forces of ethnic nationalism. It remains to be seen whether the rational view of the world encouraged by science will be strong enough to overcome this dark opposition.

Even as the powers of technology enable modern conflicts to become more destructive, the values inherent in science serve to diminish the causes leading to such conflicts. Rationality surely has a more calming influence than irrationality, and ultimately knowledge and understanding of Nature serve humanity far better than the fear mongering of ignorance and superstition.

Despite today's fashionable attacks, the concept of scientific truth has served our civilization with extraordinary success. It is not a comforting truth, but a well-known remark made by Steven Weinberg—"The more the universe seems comprehensible, the more it seems pointless"[40]—expresses a disillusionment that appears to me inappropriate: a *point* (in the sense Weinberg intended, that is, a *purpose*) is not what our truth is about. At a later time, Weinberg confessed that his statement expressed nostalgia "for a world in which the heavens declared the glory of God."[41] This longing for a fusion of emotional, religious, and scientific truths lies behind some of the contemporary attacks on the research enterprise, but a fusion of these spheres is irretrievably lost. Nevertheless, science is certainly capable of instilling great emotional satisfaction; indeed, many physicists stand in reverence before the grandeur of Nature and the beauty of its structure as revealed by their theories and discoveries. Listen to Poincaré:

> The scientist does not study nature because it is useful; he studies it because he delights in it, and he delights in it because it is beautiful. If nature were not beautiful, it would not be worth knowing, and if nature were not worth knowing, life would not be worth living.[42]

We must settle for the delight of knowing how Nature works. The various kinds of truths that have been torn asunder—scientific and religious, rational and emotional—can never be joined together again.

NOTES

INTRODUCTION

1. Laudan, *Science and Relativism*, p. x.

1 CONVENTIONS

1. Cromer, *Uncommon Sense*, p. 70.
2. Wolpert, *The Unnatural Nature of Science*, p. 47.
3. Cited in ibid., p. 48.
4. Harding, *The Science Question in Feminism*, p. 240.
5. Boyle quoted on p. 77 of Shapin and Schaffer, *Leviathan and the Air-Pump.*
6. Ibid., p. 79; Shapin and Schaffer use the phrase "form of life" from Wittgenstein, meaning essentially the same as *Weltanschauung.*
7. Ibid., p. 339.
8. Bloor, *Knowledge and Social Imagery*, p. 26.
9. Einstein and Infeld, *The Evolution of Physics*, p. 294.
10. Einstein, *Ideas and Opinions*, p. 272.
11. Einstein, *Mein Weltbild*, p. 168; my translation.
12. Poincaré, *The Foundations of Science*, p. 28.
13. Ibid., pp. 123, 124.
14. Ibid., p. 28.
15. Feigl, "Beyond peaceful coexistence," p. 9.
16. Popper, *The Logic of Scientific Discovery*, p. 82; italics in original.
17. It turned out later that some experiments done almost thirty years earlier had suggested that parity might not be conserved in beta decay, but they had been considered erroneous and were therefore ignored—everyone *knew* that parity was conserved! Those who stress the staying power of established theories even against contrary evidence have a point.

18. Bloor, *Knowledge and Social Imagery*, p. 117.
19. Evans-Pritchard, *Witchcraft, Oracles, and Magic among the Azande*, p. 24.
20. Bloor, *Knowledge and Social Imagery*, p. 124.
21. Evans-Pritchard, *Witchcraft, Oracles, and Magic among the Azande*, p. 25.
22. Post, "Introduction to a general theory of elementary propositions," pp. 163–185.
23. Reichenbach, *Philosophic Foundations of Quantum Mechanics.*
24. In his derivation of the law governing the colors of light emitted by a black body, Max Planck had to assume that light of a given frequency always came in discrete packages whose energy is proportional to that frequency; the constant of proportionality was named after him.
25. Bloor, *Knowledge and Social Imagery*, pp. 108ff.

2 SCIENCE AS A SOCIAL CONSTRUCT?

1. I am indebted to Richard S. Westfall for pointing out to me that the case of Giordano Bruno, often cited in this connection, falls into a different category.
2. Shapin and Schaffer, *Leviathan and the Air-Pump*, pp. 283, 284.
3. "Science, like every other human product, is racial and conditioned by blood." (Quoted from Lenard in Clark, *Einstein*, pp. 525–526.)
4. Harding, *The Science Question in Feminism*, p. 25; italics in original.
5. Appleyard, *Understanding the Present*, p. 76.
6. Ibid., p. 54.
7. For a relevant discussion of this connection, see Gordon, "Darwin and political economy."
8. For an analysis of the impact of metaphors and social conflicts on a prominent controversy in geology, see Rudwick, *The Great Devonian Controversy.*
9. Bridgman, *The Nature of Thermodynamics*, p. 3.
10. Harding, *The Science Question in Feminism*, p. 237.
11. Ibid., p. 113.
12. Ibid., p. 116.
13. Keller, *Secrets of Life, Secrets of Death*, p. 91.
14. Forman, "Weimar culture, causality, and quantum theory."
15. Ibid., p. 108.
16. Ibid., p. 109; italics added.
17. Ibid., p. 115; italics in original.
18. Feuer, *Einstein and the Generations of Science*, p. 170.
19. Forman, "Weimar culture, causality, and quantum theory," p. 110; italics in original.
20. Heisenberg, *Physics and Beyond*, p. 9.

21. See also Hendry, "Weimar culture and quantum causality," for arguments against Forman's thesis.
22. Shapin, *A Social History of Truth,* p. 344.
23. Bloor, *Knowledge and Social Imagery,* p. 37.
24. Collins, *Changing Order,* p. 165.
25. Ibid., p. 125.
26. Ibid., p. 74; italics in original.
27. Two instances leading to pseudo-facts will be discussed in Chapter 5.
28. The physicist Dayton Miller interpreted the outcome of his repetition of the Michelson-Morley experiment in 1925 as falsifying Einstein's theory, but by then relativity had been so well corroborated by a network of many other results that Miller's experiment was ignored.
29. Collins, *Changing Order,* p. 143.
30. Collins and Cox, "Recovering relativity," p. 439.
31. Latour and Woolgar, *Laboratory Life;* Latour, *Science in Action.*
32. Pickering, *Constructing Quarks.*
33. Ibid., p. 19, n. 13.
34. Bloor, *Knowledge and Social Imagery,* p. 85.
35. Frege, *The Foundations of Arithmetic,* p. 35.
36. Bloor, *Knowledge and Social Imagery,* p. 86.
37. Ibid., p. 87.
38. Ibid., p. 2.
39. Ibid., p. 7.
40. Bloor, "The strength of the strong program," p. 205.
41. Bloor, *Knowledge and Social Imagery,* p. 32.
42. Ibid.
43. Ibid., p. 38.
44. Chalmers, *Science and Its Fabrication,* p. 92.
45. Shapin, *A Social History of Truth,* p. 16.
46. Latour and Woolgar, *Laboratory Life,* p. 31.
47. Latour, *Science in Action,* p. 25.
48. Latour and Woolgar, *Laboratory Life,* p. 243; italics in original.
49. Ibid., p. 180; italics in original.
50. Latour, *Science in Action,* p. 99; italics in original.
51. Latour and Woolgar, *Laboratory Life,* p. 176; italics in original.
52. Ibid., p. 237.
53. Ibid., p. 182.
54. Ibid.
55. Ibid., p. 238.
56. For an elaboration of this kind of argument against constructionism, see Laudan, *Science and Relativism,* pp. 158ff.

57. Ibid., p. 257; italics in original.
58. Harding, *The Science Question in Feminism*, pp. 249–250.
59. Ibid.
60. Latour, *Science in Action*, p. 100.
61. Bruno Latour, letter to *The Sciences*, March/April 1995, p. 6; italics in original.
62. Pickering, *Constructing Quarks*, p. 413.
63. Gargamelle was the name of a very large bubble chamber at CERN, used for the detection of particles.
64. Pickering, *Constructing Quarks*, p. 405; italics in original.
65. Ibid., p. 409.
66. Ibid., p. 410.
67. Pickering, "Against putting the phenomena first," p. 87.
68. Galison, *How Experiments End*, p. 258.
69. Pickering, *Constructing Quarks*, p. 413.
70. Ibid.
71. Ibid., p. 406; italics in original.
72. Ibid., p. 413. In his most recent book, *The Mangle of Practice*, Pickering adopts a rather more cautious attitude, though he never explicitly repudiates his earlier extreme position. The closest he seems to come to his previous stance is on p. 209: "Scientific culture, then, including scientific knowledge, has to be seen as intrinsically historical, in that its specific contents are a function of the temporally emergent contingencies of its production."
73. In his Rothchild lecture of 1992, pp. 8–9, Kuhn declared: "I am among those who have found the claims of the strong program absurd: an example of deconstruction gone mad."
74. For a deliberate parody of the writings of relativistic social constructivists by a physicist, see Sokal, "Transgressing the boundaries," which was published in the journal *Social Text* (without the editors' awareness of its ironic intent). The hoax engendered a front-page story in the *New York Times* of May 18, 1996, followed by a contributed column and a number of letters.

3 THE AIM OF SCIENCE IS UNDERSTANDING

1. Bronowski, *Science and Human Values*, p. 10.
2. Weinberg, *Dreams of a Final Theory*.
3. Appleyard, *Understanding the Present*.
4. This assessment is based on my memory of Cohen's lectures almost fifty years ago, when I was a teaching assistant in one of his courses.
5. Kuhn, *The Structure of Scientific Revolutions*.

6. This is the phrase used by the science philosopher Imre Lakatos for a rather similar idea.
7. A century later, after the development of quantum mechanics and nuclear physics, Hans Bethe gave an explanation based on detailed thermonuclear processes, which showed Helmholtz's result to be only a small part of the answer.
8. Einstein, *Out of My Later Years,* p. 54
9. See Klein, "Some turns of phrase in Einstein's early papers," pp. 369–373.
10. Einstein, "Über einen die Erzeugung und Verwandlung des Lichtes betreffenden heuristischen Gesichtspunkt."
11. This is an old conundrum of mathematics: if all the theorems of arithmetic are logical consequences of the axioms, in what sense can mathematicians claim to find anything "new" in the theory of numbers? It took some 350 years before Fermat's last theorem was finally proved—surely this theorem does assert something new, even though it is a logical consequence of the axioms of arithmetic.
12. Gell-Mann, *The Quark and the Jaguar,* pp. 99–100.
13. A linear equation, as in algebra, contains the unknown only in the first power; a nonlinear equation may contain it quadratically, or in some other form. Readers who are unfamiliar with the concepts of differentiation and differential equations may find simple and brief introductions to these ideas in my book, *What Makes Nature Tick?*
14. For examples, see Hadamard, *The Psychology of Invention in the Mathematical Field,* pp. 116–123.
15. Kane, *The Particle Garden,* p. 166.
16. Weinberg, *Dreams of a Final Theory,* p. 55.
17. Quoted by Weinberg, *Dreams of a Final Theory,* p. 61.
18. Ibid., p. 57; italics in original.
19. In some contexts *reductionism* refers only to certain specific kinds of reductions, and its opposite does not necessarily mean that some phenomena are *sui generis.* I am not here referring to this meaning.

4 EXPLANATORY DEVICES

1. Kauffman, *The Origins of Order: Self-Organization and Selection in Evolution.*
2. *Nature,* 373 (1995), p. 555.
3. Holyoak and Thagard, *Mental Leaps.*
4. Holton, *Einstein, History, and Other Passions,* pp. 94–95.
5. This word does not seem to exist, but it should: it means the biological history of life.
6. Recently, however, some cosmologists have argued that there was no choice in the initial state of the universe if it was, indeed, a singularity.
7. Carter, "Large number coincidences and the anthropic principle in cos-

mology"; Barrow and Tipler, *The Anthropic Cosmological Principle;* Gale, "The anthropic principle."

8. Collins and Hawking, "Why is the universe isotropic?"
9. Corey, *God and the New Cosmology.*
10. Barrow and Tipler, *The Anthropic Cosmological Principle,* p. 16.
11. See, for example, Gardner, "WAP, SAP, PAPA, and FAP."

5 THE ROLE OF FACTS

1. Poincaré, *The Foundations of Science,* p. 128.
2. It is an interesting historical fact that when a spectral analysis was first performed on the light from the sun, one spectrum was found that had no Earthly analogue. Only later was the element that emits this spectrum discovered on Earth. It was named helium, after the Greek name for the sun.
3. Holton, *Einstein, History, and Other Passions,* p. 65.
4. Such classifications were at one time regarded as explanations in Plato's sense of finding the ideal forms of organisms and the relations between them. We no longer give such arguments any explanatory value.
5. For a description of the details of this discovery and the ensuing controversy, see Holton, *The Scientific Imagination,* pp. 25–83.
6. For details, see Nye, "N-rays: An episode in the history and psychology of science."
7. See Allen, "The rise and fall of polywater."
8. For a detailed description of this controversy, see Collins, *Changing Order,* pp. 79–112.
9. For detailed descriptions both of the Weber affair and the cold-fusion fiasco, see Collins and Pinch, *The Golem,* Chapters 3 and 5.
10. Descartes had declared all of his philosophy would be null and void if the speed of light were not infinite.
11. We may note parenthetically that Hubble's discovery of this linear relation, which has been confirmed by many later measurements, was a coup of intuition. It required a great deal of imagination to read his observational data as indicating simple proportionality.
12. This is determined by the fact that it *appears* to us to lie inside it. If a stellar object is seen as apparently belonging to a large group of stars, it is much more likely to belong to it, in fact, than to be far closer or much further away, appearing to be part of the group only by happenstance.
13. See, for example, John Maddox, "More muddle over the Hubble constant"; Bolte and Hogan, "Conflict over the age of the universe."
14. See Chalmers, *Science and Its Fabrication,* pp. 64f.
15. Poincaré, *The Foundations of Science,* p. 129.

6 THE BIRTH AND DEATH OF THEORIES

1. I disagree with the elaborate argument advanced by Paul Feyerabend (*Against Method*, p. 152) "abolishing the distinction between a context of discovery and a context of justification." Thomas Kuhn, in *The Essential Tension*, pp. 326–327, similarly denies the importance of this distinction.
2. Holton, *Einstein, History, and Other Passions*, p. 111.
3. On some later occasions Einstein denied that he knew about it. In the second paragraph of the 1905 paper in which he proposed the theory of relativity, he refers to "unsuccessful attempts to discover a motion of the Earth with respect to the 'light medium'," but there had been a number of other such experiments. See Holton, *Thematic Origins of Scientific Thought*, Chapter 8 and pp. 477–480.
4. Holton, *Einstein, History, and Other Passions*, p. 63.
5. For a historical study in the case of genetics in Germany, see Harwood, *Styles of Scientific Thought*.
6. Hadamard, *The Psychology of Invention in the Mathematical Field*, p. 31.
7. Quoted in Mehra, *The Beat of a Different Drum*, p. 494; italics in original.
8. Hadamard, *Psychology of Invention in the Mathematical Field*, p. 8.
9. Ellis, *Truth and Objectivity*, p. 6.
10. However, Cushing argues in *Quantum Mechanics: Historical Contingency and the Copenhagen Hegemony* that a theory by David Bohm is a viable alternative to the generally accepted quantum theory, and experimentally indistinguishable from it. I shall return to this theory later on.
11. Holton, *Einstein, History, and Other Passions*, pp. 97ff.
12. Peterson, *Newton's Clock: Chaos in the Solar System*.
13. Gellner, *Legitimation of Belief*, p. 176.
14. Lakatos, "Falsification and the methodology of scientic research programmes," p. 142; italics in original.
15. Ibid., p. 121; italics in original.
16. Ibid., p. 119; italics in original.
17. Ibid., p. 107; italics in original.
18. Wolpert, *The Unnatural Nature of Science*, p. 95.
19. Polkinghorne, *Rochester Roundabout*, p. 171.
20. Max Planck, lecture on September 24, 1891, in Halle an der Saale, quoted in Jost, *Das Märchen vom Elfenbeinernen Trum*, p. 75; my translation.
21. Laudan, *Beyond Positivism and Relativism*, p. 21.
22. Duhem, *The Aim and Structure of Physical Theory*, pp. 188ff.
23. Actually, this spot, which is now called "Poisson's bright spot," was first seen more than fifty years earlier by Giovanni Domenico Maraldi (known as Maraldi II), but it was forgotten. See Strong, *Concepts of Classical Optics*, p. 186. Although Strong does not specify which Maraldi saw the spot, if

his timing is correct it must have been Maraldi II, because his uncle, Maraldi I, had died 89 years before the Poisson controversy.

24. Ellis, *Truth and Objectivity,* p. 108.
25. For some interesting instances of "crucial experiments" and their interpretations, see Collins and Pinch, *The Golem,* Chapter 2.
26. Feyerabend, *Against Method,* p. 14; italics in original.
27. Ibid., p. 1; italics in original.
28. Ziman, *Reliable Knowledge.*
29. Polanyi, *Personal Knowledge,* p. 163.
30. See Rudwick, *The Great Devonian Controversy.*
31. There are, however, some phenomena at the everyday level whose explanation requires the quantum theory: phase transitions, superconductivity, and other such collective effects.

7 THE POWER OF MATHEMATICS

1. Poincaré, *The Foundations of Science,* p. 130.
2. Laplace, *Essai sur les probabilités,* p. 4.
3. Unless the arguments, mentioned in Chapter 4, which purport to be able to deduce the initial state of the universe from the nature of its evolution equations and the existence of a "big bang" singularity, are correct.
4. Poincaré, *The Foundations of Science,* p. 134.
5. Thereby hangs an interesting tale. After Poincaré had won a prize from the King of Sweden for a paper on celestial mechanics, he found an error in his work, the correction of which led him to the discovery of what is now called chaotic behavior. Since he could not repair the error in time, he spent all of the prize money buying back the copies of the journal containing his erroneous article. See Diacu and Holmes, *Celestial Encounters.*
6. The more accurate word applicable here is *ergodic.*
7. Laskar, "Large scale chaos and marginal stability in the solar system."
8. Einstein, *Ideas and Opinions,* p. 274.
9. Poincaré, *The Foundations of Science,* pp. 348–349.
10. Weinberg, *Dreams of a Final Theory,* p. 56.
11. Ziman, *Reliable Knowledge;* italics in original.
12. Wigner, "The unreasonable effectiveness of mathematics in the natural sciences," p. 223.
13. Hardy, *A Mathematician's Apology,* pp. 123–124.
14. "Intuitionism" received its name from adherents who believe that the concept of *number* is based on an immediate human intuition and is irreducible. This stands in contrast to the ideas of those, like Bertrand Rus-

sell and Alfred North Whitehead, who want to ground that concept on logic and the theory of sets.

15. Hilbert, *Gesammelte Abhandlungen*, vol. 3, p. 159.
16. The issue of standards of mathematical rigor is somewhat contentious within mathematics and subject to historical flux, as noted in Chapter 2.
17. Dyson, "Mathematics in the physical sciences," p. 106; emphasis added.

8 CAUSALITY, DETERMINISM, AND PROBABILITY

1. Popper, *The Logic of Scientific Discovery*, p. 247.
2. See, for example, Galison, *How Experiments End*, p. 266.
3. This is the crux of the plot of the movie *Terminator 2*, which, however, ignores the resulting problems.
4. My discussion of statistical mechanics in Chapter 3 led to the conclusion that the direction of this time-arrow defined by causality also determines the direction of the time-arrow defined by the second law of thermodynamics. The latter comes about entirely from the way all experimental questions are posed: if we arrange circumstances A at the time t_1, what will be observed at the *later* time t_2? It is the cause A that we control, and hence A occurs earlier. If the world were so arranged that effects preceded their causes, entropy would always decrease. In that case, presumably, we would psychologically experience the flow of time in the opposite direction, and Nature would appear just the same as now. Thus the psychological arrow of time is also determined by the arrow of causality, and it would be inconceivable for aliens on a faraway planet to experience time running in the opposite direction.
5. An inertial frame is a frame of reference or laboratory in which the Newtonian equations of motion hold in their original form. Any reference frame moving with uniform velocity with respect to an inertial frame is also an inertial frame; an accelerated laboratory is not.
6. For a detailed explanation of how that could be done, see Newton, *What Makes Nature Tick?*, pp. 137ff.
7. If it is not specified so precisely, it is described by a *density matrix*. We need not pay attention to this subtlety, however, or to the meaning of these words.
8. There are certain mathematical difficulties associated with the need for infinite sequences in the frequency theory, but we may ignore these here.
9. Assuming that both a and b are very large, as they should be for the calculation of the probability.
10. Popper, *Realism and the Aim of Science*, p. 287; italics in original.
11. Ibid., p. 286.

12. Ibid., p. 282n.
13. Ibid., p. 351; italics in original.
14. Ibid., p. 359; italics in original.
15. Ibid., p. 397; italics in original.
16. Gell-Mann and Hartle, "Quantum mechanics in the light of quantum cosmology," pp. 425–458; Omnès, "Consistent interpretations of quantum mechanics"; Omnès, *The Interpretation of Quantum Mechanics.*
17. Popper, *Realism and the Aim of Science,* p. 398.

9 REALITY ON TWO SCALES

1. Einstein, *Mein Weltbild,* p. 208; my translation.
2. According to Plato.
3. Chalmers, *Science and Its Fabrication,* pp. 52ff.
4. Letter to Richard Phillips, April 15, 1846, in Faraday, *Experimental Researches in Electricity,* vol. 3, pp. 885ff.
5. Einstein, *Mein Weltbild,* p. 213; my translation.
6. Abdus Salam, in Taylor, *Tributes to Paul Dirac,* p. 90.
7. Hacking, *Representing and Intervening,* pp. 22–23; italics in original.
8. This apparent contradiction is entirely accounted for by a consequence of the relativistic transformation law: namely, that the simultaneity of two events at different locations depends on the motion of the observer. I claim your clock is slow because, as it moves past a string of synchronous clocks at rest in my laboratory, it keeps falling behind. You agree with that observation, but your explanation is that my clocks are not synchronous. I say your stick is shorter than mine because as it passes, I mark off its ends simultaneously and find the distance between the marks shorter than my stick; your explanation is that my marks were not made simultaneously.
9. See Gell-Mann, *The Quark and the Jaguar;* Gell-Mann and Hartle, "Quantum mechanics in the light of quantum cosmology," pp. 425–458; also Omnès, *The Interpretation of Quantum Mechanics.*
10. Actually, this controvery was caused by more than poor statistics. The CERN group, it seems, had committed a basic error in scientific methodology: it had treated batches of data that showed the dip in the curve differently from batches that did not, scrutinizing the latter more carefully for "something wrong" and always finding it. See Cromer, *Uncommon Sense,* pp. 169–170.
11. The wave nature of light implies that a light signal of a single wavelength must be infinitely long; a short signal is necessarily a superposition of many different wavelengths. No light signal shorter than 10^{-15} seconds can have an identifiable color.

12. Quoted by Aage Petersen, "The philosophy of Niels Bohr," *Bulletin of the Atomic Scientists*, 19 (September 1963), pp. 10–11.
13. Ibid., p. 12; italics in original.
14. Heisenberg, *Physics and Philosophy*, p. 144.
15. Ibid., p. 129.
16. Ibid., p. 186.
17. Ibid., p. 145.
18. Quoted by Aage Petersen, "The philosophy of Niels Bohr," p. 12.
19. Heisenberg, *Physics and Philosophy*, p. 56.

10 REALITY AT THE SUBMICROSCOPIC LEVEL

1. This difficulty with a "realistic" interpretation of the wave function was pointed out to Schrödinger by the great Dutch physicist Hendrik Lorentz in a letter dated May 27, 1926 (see Przibram, *Letters on Wave Mechanics*, p. 44), and in his reply, dated June 6, 1926, Schrödinger indicates that he had been aware of this problem for some time (ibid., pp. 55f.).
2. More precisely, the square of the absolute magnitude of the wave function, which in general is a quantity that is complex in the mathematical sense, multiplied by the infinitesimal volume element is the probability of finding the particle in that volume element. For simplicity, we are assuming that the state of the system is as completely specified as quantum mechanics permits.
3. This is why, in his mathematical formulation of the quantum theory, John von Neumann equated measurement with the preparation of a state.
4. It has to be described by a density matrix.
5. Technically speaking, the entanglement is a definite phase relation between the two states.
6. Popper, *Quantum Theory and the Schism in Physics*, pp. 72–74.
7. Monroe et al., "A 'Schrödinger cat' superposition state of an atom."
8. We shall ignore here the question of whether it makes sense at all to describe a vastly complicated system in [illegible] roundings, like a living cat, for [illegible]
9. From *The Born-Einste*[illegible] min in "Is th[illegible]
10. Einstei[illegible]

ability for each of them: what is the probability for finding particle 1 with property *a*, assuming that particle 2 has property *b*? If this probability depends on *b*, the two are entangled. The probability of the outcome of any observation to ascertain if particle 1 has property *a* will then necessarily depend on the outcome of an observation on particle 2, which may seem like "spooky action at a distance" and raise the question "How does particle 1 know about particle 2?" Only in the special case in which the joint probability happens to be the product of individual probabilities will the probability of *a* for particle 1 not depend on the property of particle 2; in that case, the two are independent rather than entangled.

13. Bohm and Hiley, *The Undivided Universe.*
14. Cushing, *Quantum Mechanics.*
15. This version is due to Mermin, "Is the moon there?" pp. 38–47.
16. I disagree with Paul Teller's confused claim, in *An Interpretive Introduction to Quantum Field Theory,* pp. 97ff., that the quantum field cannot be interpreted as an *operator-valued* field function of space-time.
17. Quoted in Schweber, *QED and the Men Who Made It,* pp. 355ff., from an unpublished public lecture by Schwinger in the early 1960s, pp. 62–64.
18. Heisenberg, *Physics and Philosophy,* p. 143.
19. This word was much used by Heisenberg in the 1920s, but during the 1930s he modified his public calls for dispensing with it, possibly for political reasons.
20. From Schwinger's unpublished lecture, p. 73.

11 TRUTH AND OBJECTIVITY

1. In his extensive article "Consistent interpretations of quantum mechanics," Roland Omnès claims "the result of an experiment is always true" and repeatedly refers to the "truth of a property." I confess I do not quite understand what he means by this.
2. James, *Pragmatism and The Meaning of Truth,* p. 108; italics in original.
3. An example is the large role played by the word *existence* in parts of Western philosophy. These philosophical discussions are untranslatable and incomprehensible in some languages in which there is no analogous noun, which says more about the depth of the debates than about the languages.
4. Tarski, *Logic, Semantics, Meta-Mathematics.*
5. The confusion of the two meanings of this word, the ethical and epistemological meaning on one hand and that of Einstein's theory on the other, led to much initial hostility toward the latter. I trust that the meaning of the word I am using in this chapter will not be confounded with Einstein's.

6. Latour and Woolgar, *Laboratory Life,* p. 243; italics in original.
7. Popper, "Normal science and its dangers," p. 56.
8. Rorty, *Philosophy and the Mirror of Nature.*
9. Hacking, *Representing and Intervening.*
10. Ziman, *Reliable Knowledge,* p. 107.
11. Fine, *The Shaky Game,* pp. 140–141.
12. Kuhn, *The Structure of Scientific Revolutions.*
13. Shimony, *Search for a Naturalistic World View,* vol. 1, p. 310.
14. Quoted in Bronowski, *Science and Human Values,* p. 30.
15. James, *Pragmatism and The Meaning of Truth,* p. 271.
16. Bronowski, *Science and Human Values,* p. 52.
17. Keller, *Secrets of Life, Secrets of Death,* p. 92.
18. Ziman, *Reliable Knowledge,* p. 83; italics in original.
19. Poincaré, *The Foundations of Science,* p. 367.
20. Cushing, *Quantum Mechanics.*
21. Poincaré, *The Foundations of Science,* p. 352. He was referring to classifications, but what he said applies equally well to what we are talking about here.
22. Elliott Lieb emphasizes the universal transcultural, nature of the language of mathematics, which enhances its value as a vehicle for transmitting scientific truth (private communication).
23. Quoted in Rauch, *Kindly Inquisitors,* pp. 36–37, from Raymond A. Moody, Jr. *Elvis after Life: Unusual Psychic Experiences Surrounding the Death of a Superstar* (Atlanta: Peachtree, 1988), excerpted in *Harper's,* August 1988.
24. From a letter to Lady Welby of December 23, 1908; quoted in Rauch, *Kindly Inquisitors,* p. 56; italics in original.
25. Chalmers, *Science and Its Fabrication,* p. 49; italics in original.
26. Contrary to some popular contentions, instances of scientific dishonesty or fraud are extremely rare and confined almost entirely to areas near or at the border of practical applications, where monetary rewards play a larger role, rather than in basic science itself. Admittedly, though, hard data on this topic are almost nonexistent.
27. Rauch, *Kindly Inquisitors,* p. 67; italics in original.
28. Ibid.; italics in original.
29. Hubbard, "Science, Facts, and Feminism," p. 125.
30. Ibid., p. 128.
31. Ibid.
32. Poincaré, *The Foundations of Science,* p. 355.
33. Quoted from the *Evanston Colloquium,* p. 46, in Hadamard, *The Psychology of Invention,* p. 107.
34. Quoted from *Revue des Deux Mondes,* January–February 1915, p. 657, in Hadamard, *The Psychology of Invention,* p. 107.

35. Duhem, *The Aim and Structure of Physical Theory,* pp. 55ff.
36. Peirce, *Collected Papers,* vol. VII, p. 56.
37. Quoted in Schweber, *QED and the Men Who Made It,* p. 463, from a public address by Feynman.
38. My translation from the polemical essay "Eine Duplik." See Gotthold Ephraim Lessing, *Sämmtliche Schriften* (Berlin, 1839), vol. X, pp. 49–50.
39. Appleyard, *Understanding the Present.*
40. Weinberg, *The First Three Minutes,* p. 154.
41. Weinberg, *Dreams of a Final Theory,* p. 256.
42. Poincaré, *The Foundations of Science,* p. 366.

FURTHER READING

GENERAL

Cromer, Alan. *Uncommon Sense: The Heretical Nature of Science.* Oxford: Oxford University Press, 1993.

Einstein, Albert. *Ideas and Opinions.* New York: Crown Publishers, 1954.

Hempel, Carl G. *Aspects of Scientific Explanation, and Other Essays in the Philosophy of Science.* New York: Free Press, 1965.

Holton, Gerald, and Robert S. Morison, eds. *Limits of Scientific Inquiry.* New York: W. W. Norton, 1979.

Laudan, Larry. *Beyond Positivism and Relativism: Theory, Method, and Evidence.* Boulder, Colo.: Westview, 1996.

Medawar, Peter B. *The Art of the Soluble.* London: Methuen, 1967.

Weinberg, Steven. *Dreams of a Final Theory.* New York: Pantheon, 1992.

Wolpert, Lewis. *The Unnatural Nature of Science.* Cambridge, Mass.: Harvard University Press, 1992.

1 CONVENTIONS

Poincaré, Henri. *The Foundations of Science (Science and Hypothesis, The Value of Science, Science and Method).* Lancaster, Penn.: Science Press, 1946.

Popper, Karl R. *The Logic of Scientific Discovery.* New York: Basic Books, 1959. [Sections 11, 19, and 46.]

2 SCIENCE AS A SOCIAL CONSTRUCT?

Chalmers, Alan F. *What Is This Thing Called Science? An Assessment of the Nature and Status of Science and Its Methods,* 2d ed. Milton Keynes: Open University Press, 1982.

——— *Science and Its Fabrication.* Minneapolis: University of Minnesota Press, 1990.

Cole, Stephen. *Making Science: Between Nature and Society.* Cambridge, Mass.: Harvard University Press, 1992.

Gross, Paul R., and Norman Levitt. *Higher Superstition: The Academic Left and Its Quarrels with Science.* Baltimore: Johns Hopkins University Press, 1994.

Gross, Paul R., Norman Levitt, and Martin W. Lewis, eds. *The Flight from Science and Reason.* New York: New York Academy of Science, 1996.

Holton, Gerald. *Science and Anti-Science.* Cambridge, Mass.: Harvard University Press, 1993.

Laudan, Larry. *Science and Relativism: Some Key Controversies in the Philosophy of Science.* Chicago: University of Chicago Press, 1996.

Rauch, Jonathan. *Kindly Inquisitors: The New Attacks on Free Thought.* Chicago: University of Chicago Press, 1993.

Ziman, John. *Reliable Knowledge: An Exploration of the Grounds for Belief in Science.* Cambridge: Cambridge University Press, 1978.

3 THE AIM OF SCIENCE IS UNDERSTANDING

Braithwaite, Richard Bevan. *Scientific Explanation: A Study of the Function of Theory, Probability and Law in Science.* Cambridge: Cambridge University Press, 1964.

Duhem, Pierre. *The Aim and Structure of Physical Theory.* Princeton: Princeton University Press, 1991.

Feynman, Richard. *The Character of Physical Law.* Cambridge, Mass.: MIT Press, 1993.

Holton, Gerald. *The Scientific Imagination: Case Studies.* Cambridge: Cambridge University Press, 1978.

4 EXPLANATORY DEVICES

Barrow, John D., and Frank J. Tipler. *The Anthropic Cosmological Principle.* Oxford: Oxford University Press, 1986.

Brown, James Robert. *The Laboratory of the Mind: Thought Experiments in the Natural Sciences.* London: Routledge, 1991.

Carter, Brandon. "Large number coincidences and the anthropic principle in cosmology." In *Confrontation of Cosmological Theories with Observation,* ed. M. S. Longhair, p. 291. Dordrecht: Reidel, 1974.

Corey, M. A. *God and the New Cosmology: The Anthropic Design Argument.* Lanham, Md.: Rowman and Littlefield, 1993.

Crick, Francis. *Life Itself: Its Origin and Nature.* New York: Simon and Schuster, 1981.

Dawkins, Richard. *The Blind Watchmaker: Why the Evidence of Evolution Reveals a Universe without Design.* New York: Norton, 1987.

Gale, George. "The anthropic principle." *Scientific American,* December 1981, p. 154.

Gardner, Martin. "WAP, SAP, PAPA, and FAP." *The New York Review of Books,* 33 (May 8, 1986), pp. 22–25.

Holyoak, Keith J., and Paul Thagard. *Mental Leaps: Analogy in Creative Thought.* Cambridge, Mass.: MIT Press, 1995.

Judson, Horace F. *The Eighth Day of Creation: The Makers of the Revolution in Biology.* New York: Simon and Schuster, 1979.

Sorensen, Roy A. *Thought Experiments.* Oxford: Oxford University Press, 1992.

5 THE ROLE OF FACTS

Allen, Leland. "The rise and fall of polywater." *New Scientist,* 59 (1973), pp. 376–380.

Bolte, M., and C. J. Hogan. "Conflict over the age of the universe." *Nature,* 376 (August 3, 1995), pp. 399–402.

Collins, H. M. *Changing Order: Replication and Induction in Scientific Practice.* London: Sage, 1985. [Pages 78ff., concerning gravitational waves.]

——— "Son of seven sexes: The social destruction of a physical phenomenon." *Social Studies of Science,* 11 (1981), pp. 33–62. [Concerning gravitational waves.]

Huizinga, John R. *Cold Fusion: The Scientific Fiasco of the Century.* Rochester, N.Y.: University of Rochester Press, 1992.

Kragh, Helge. *Cosmology and Controversy: The Historical Development of Two Theories of the Universe.* Princeton: Princeton University Press, 1996.

Nye, Mary-Jo. "N-rays: An episode in the history and psychology of science." *Historical Studies in the Physical Sciences,* 11 (1980), pp. 125–156.

6 THE BIRTH AND DEATH OF THEORIES

Lakatos, Imre. "Falsification and the methodology of scientific research programmes." In *Criticism and the Growth of Knowledge,* ed. Imre Lakatos and Alan Musgrave, pp. 91–195. Cambridge: Cambridge University Press, 1970.

Popper, Karl R. *The Logic of Scientific Discovery.* New York: Basic Books, 1959.

7 THE POWER OF MATHEMATICS

Barrow, John D. *Pi in the Sky: Counting, Thinking and Being.* Oxford: Clarendon Press, 1992.

Dyson, Freeman. "Mathematics in the physical sciences." In *The Mathematical Sciences,* ed. Committee on Support of Research in the Mathematical Sciences, pp. 97–115. Cambridge, Mass.: MIT Press, 1969.

Hardy, G. H. *A Mathematician's Apology.* Cambridge: Cambridge University Press, 1993.

Wigner, Eugene. "The unreasonable effectiveness of mathematics in the natural sciences." In *Symmetries and Reflections,* pp. 222–237. Bloomington: Indiana University Press, 1967.

8 CAUSALITY, DETERMINISM, AND PROBABILITY

Popper, Karl. *Realism and the Aim of Science.* Totowa, N.J.: Rowman and Littlefield, 1983.

9 REALITY ON TWO SCALES

Heisenberg, Werner. *Physics and Philosophy: The Revolution in Modern Science.* New York: Harper, 1958.

10 REALITY AT THE SUBMICROSCOPIC LEVEL

Cushing, James T. *Quantum Mechanics: Historical Contingency and the Copenhagen Hegemony.* Chicago: University of Chicago Press, 1994.

d'Espagnat, Bernard. *Veiled Reality: An Analysis of Present-Day Quantum Mechanical Concepts.* Reading, Mass.: Addison-Wesley, 1995.

Heisenberg, Werner. *Physics and Beyond: Encounters and Conversations.* New York: Harper and Row, 1971.

——— *Physics and Philosophy: The Revolution in Modern Science.* New York: Harper, 1958.

Jammer, Max, *The Philosophy of Quantum Mechanics: The Interpretation of Quantum Mechanics in Historical Perspective.* New York: Wiley, 1974.

Mermin, N. David. "Is the moon there when nobody looks? Reality and the quantum theory." *Physics Today,* April 1985, pp. 38–47.

Omnès, Roland. *The Interpretation of Quantum Mechanics.* Princeton: Princeton University Press, 1994.

11 TRUTH AND OBJECTIVITY

Bronowski, Jacob. *Science and Human Values.* New York: Harper and Row, 1965.

Ellis, Brian. *Truth and Objectivity.* Oxford: Blackwell, 1990.

Graham, Loren R. *Between Science and Values.* New York: Columbia University Press, 1981.

Horwich, Paul. *Truth.* Oxford: Blackwell, 1990.

Rescher, Nicholas. *The Coherence Theory of Truth.* Oxford: Clarendon Press, 1973.

BIBLIOGRAPHY

Allen, Leland. "The rise and fall of polywater." *New Scientist*, 59 (1973), pp. 376–380.

Appleyard, Bryan. *Understanding the Present: Science and the Soul of Modern Man*. New York: Doubleday, 1993.

Azzouni, Jody. *Mathematical Myths, Mathematical Practice: The Ontology and Epistemology of the Exact Sciences*. Cambridge: Cambridge University Press, 1994.

Barker, Lewis M. *Learning and Behavior: A Psychological Perspective*. New York: Macmillan, 1994.

Barnes, Barry. *T. S. Kuhn and Social Science*. London: Macmillan, 1982.

Barrow, John D. *Pi in the Sky: Counting, Thinking and Being*. Oxford: Clarendon Press, 1992.

Barrow, John D., and Frank J. Tipler. *The Anthropic Cosmological Principle*. Oxford: Oxford University Press, 1986.

Bloor, David. *Knowledge and Social Imagery*. London: Routledge, 1976.

——— "The strength of the strong programme." *Philosophy of the Social Sciences*, 11 (1981), pp. 199–213.

Blume, Stuart S. *Toward a Political Sociology of Science*. New York: Free Press, 1974.

Bohm, D., and B. J. Hiley. *The Undivided Universe: An Ontological Interpretation of the Quantum Theory*. New York: Routledge, 1993.

Bohr, Niels. "Can quantum mechanical description of physical reality be considered complete?" *Physical Review*, 48 (1935), p. 696.

Bolte, M., and C. J. Hogan. "Conflict over the age of the universe." *Nature*, 376 (3 August 1995), pp. 399–402.

Braithwaite, Richard Bevan. *Scientific Explanation: A Study of the Function of*

Theory, Probability and Law in Science. Cambridge: Cambridge University Press, 1964.

Bridgman, Percy W. *The Nature of Thermodynamics.* Cambridge, Mass.: Harvard University Press, 1941.

Bronowski, Jacob. *Science and Human Values.* New York: Harper and Row, 1965.

Brown, James Robert. *The Laboratory of the Mind: Thought Experiments in the Natural Sciences.* London: Routledge, 1991.

Butts, Robert E., and James Robert Brown, eds. *Constructivism and Science: Essays in Recent German Philosophy.* Dordrecht: Kluwer, 1989.

Carter, Brandon. "Large number coincidences and the anthropic principle in cosmology." In *Confrontation of Cosmological Theories with Observation,* ed. M. S. Longhair, p. 291. Dordrecht: Reidel, 1974.

Chalmers, Alan F. *Science and Its Fabrication.* Minneapolis: University of Minnesota Press, 1990.

——— *What Is This Thing Called Science? An Assessment of the Nature and Status of Science and Its Methods,* 2d ed. Milton Keynes: Open University Press, 1982.

Clark, Ronald. *Einstein: The Life and Times.* New York: World Publishing, 1971.

Cole, Stephen. *Making Science: Between Nature and Society.* Cambridge, Mass.: Harvard University Press, 1992.

Collins, C. B., and S. W. Hawking. "Why is the universe isotropic?" *Astrophysical Journal,* 180 (1973), pp. 317–334.

Collins, H. M. *Changing Order: Replication and Induction in Scientific Practice.* London: Sage, 1985.

——— "Son of seven sexes: The social destruction of a physical phenomenon." *Social Studies of Science,* 11 (1981), pp. 33–62.

Collins, H. M., and G. Cox. "Recovering relativity: Did prophecy fail?" *Social Studies of Science,* 6 (1976), pp. 423–444.

Collins, H. M., and T. Pinch. *The Golem: What Everyone Should Know about Science.* Cambridge: Cambridge University Press, 1993.

Cook, Sir Alan. *The Observational Foundations of Physics.* Cambridge: Cambridge University Press, 1994

Corey, M. A. *God and the New Cosmology: The Anthropic Design Argument.* Lanham, Md.: Rowman & Littlefield, 1993.

Cox, R. T., C. G. McIlwraith, and B. Kurrelmeyer. "Apparent evidence of polarization in a beam of ß-rays." *Proceedings of the National Academy of Sciences,* 14 (1928), p. 544.

Crick, Francis. *Life Itself: Its Origin and Nature.* New York: Simon and Schuster, 1981.

Cromer, Alan. *Uncommon Sense: The Heretical Nature of Science.* Oxford: Oxford University Press, 1993.

Cushing, James T. *Quantum Mechanics: Historical Contingency and the Copenhagen Hegemony.* Chicago: University of Chicago Press, 1994.

Dawkins, Richard. *The Blind Watchmaker: Why the Evidence of Evolution Reveals a Universe without Design.* New York: Norton, 1987.

d'Espagnat, Bernard. *Veiled Reality: An Analysis of Present-Day Quantum Mechanical Concepts.* Reading, Mass.: Addison-Wesley, 1995.

Diacu, Florin, and Philip Holmes. *Celestial Encounters: The Origins of Chaos and Stability.* Princeton: Princeton University Press, 1996.

Diamond, Jared M. "Daisy gives an evolutionary answer." *Nature,* 380 (14 March 1996), pp. 103–104.

Duhem, Pierre. *The Aim and Structure of Physical Theory.* Princeton: Princeton University Press, 1991.

Dyson, Freeman. "Mathematics in the physical sciences." In *The Mathematical Sciences,* ed. Committee on Support of Research in the Mathematical Sciences, pp. 97–115. Cambridge, Mass.: MIT Press, 1969.

Einstein, Albert. *Ideas and Opinions* [translation of *Mein Weltbild*]. New York: Crown Publishers, 1954.

——— *Mein Weltbild.* Amsterdam: Querido Verlag, 1934.

——— *Out of My Later Years.* New York: Philosophical Library, 1950.

——— "Über einen die Erzeugung und Verwandlung des Lichtes betreffenden heuristischen Gesichtspunkt." *Annalen der Physik,* 17 (1905), pp. 132–148.

Einstein, Albert, and Leopold Infeld. *The Evolution of Physics: The Growth of Ideas from Early Concepts to Relativity and Quanta.* New York: Simon and Schuster, 1961.

Einstein, A., B. Podolsky, and N. Rosen. "Can quantum mechanical description of physical reality be considered complete?" *Physical Review,* 47 (1935), p. 777.

Ellis, Brian. *Truth and Objectivity.* Oxford: Blackwell, 1990.

Elvee, Richard Q., ed. *The End of Science?* Lanham, Md.: University Press of America, 1992.

Evans-Pritchard, Edward Evan. *Witchcraft, Oracles, and Magic among the Azande.* Oxford: Clarendon Press, 1937.

Faraday, Michael. *Experimental Researches in Electricity,* vol. 3. London, 1855.

Feigl, Herbert. "Beyond peaceful coexistence." In *Minnesota Studies in the Philosophy of Science,* 5 (1970), pp. 3–12.

Ferguson, Harvie. *The Science of Pleasure: Cosmos and Psyche in the Bourgeois World View.* London: Routledge, 1990.

Scientific Inquiry, ed. Gerald Holton and Robert S. Morison, pp. 227–241. New York: W. W. Norton, 1979.

——— *Science and Anti-Science.* Cambridge, Mass.: Harvard University Press, 1993.

——— *The Scientific Imagination: Case Studies.* Cambridge: Cambridge University Press, 1978.

——— *Thematic Origins of Scientific Thought—Kepler to Einstein.* Cambridge, Mass.: Harvard University Press, 1988.

——— "Thematic presuppositions and the direction of scientific advance." In *Scientific Explanation,* ed. A. F. Heath, pp. 1–27. Oxford: Clarendon Press, 1981.

Holton, Gerald, and Robert S. Morison, eds. *Limits of Scientific Inquiry.* New York: W. W. Norton, 1979.

Holyoak, Keith J., and Paul Thagard. *Mental Leaps: Analogy in Creative Thought.* Cambridge, Mass.: MIT Press, 1995.

Home, D., and M. A. B. Whitaker. "Ensemble interpretations of quantum mechanics: A modern perspective." *Physics Report,* 210 (1992), pp. 225–317.

Horowitz, Tamara, and Gerald J. Massey, eds. *Thought Experiments in Science and Philosophy.* Savage, Md.: Rowman and Littlefield, 1991.

Horwich, Paul. *Truth.* Oxford: Blackwell, 1990.

Hubbard, Ruth. "Science, facts, and feminism." In *Feminism and Science,* ed. Nancy Tuana, pp. 119–131. Bloomington: Indiana University Press, 1989.

Huizinga, John R. *Cold Fusion: The Scientific Fiasco of the Century.* Rochester, N.Y.: University of Rochester Press, 1992.

Hull, David. *Science as Process.* Chicago: University of Chicago Press, 1988.

Humphreys, Paul. "Why propensities cannot be probabilities." *The Philosophical Review,* 94 (1985), p. 557.

James, William. *Pragmatism and The Meaning of Truth.* Cambridge, Mass.: Harvard University Press, 1975.

Jammer, Max. *The Philosophy of Quantum Mechanics: The Interpretation of Quantum Mechanics in Historical Perspective.* New York: Wiley, 1974.

Jost, Res. *Das Märchen vom Elfenbeinernen Turm.* Berlin: Springer-Verlag, 1995.

Judson, Horace F. *The Eighth Day of Creation: The Makers of the Revolution in Biology.* New York: Simon and Schuster, 1979.

Kane, Gordon. *The Particle Garden: Our Universe as Understood by Particle Physicists.* Reading, Mass.: Addison-Wesley, 1995.

Kauffman, Stuart. *The Origins of Order: Self-Organization and Selection in Evolution.* Oxford: Oxford University Press, 1993.

Keller, Evelyn Fox. *Secrets of Life, Secrets of Death: Essays on Language, Gender and Science.* New York: Routledge, 1992.

Klein, Martin. "Some turns of phrase in Einstein's early papers." In *Physics as Natural Philosophy,* ed. A. Shimony and H. Feshbach, pp. 364–375. Cambridge, Mass.: MIT Press, 1982.

Knorr-Cetina, Karin D. *The Manufacture of Knowledge: An Essay on the Constructivist and Contextual Nature of Science.* New York: Pergamon, 1981.

Kragh, Helge. *Cosmology and Controversy: The Historical Development of Two Theories of the Universe.* Princeton: Princeton University Press, 1996.

Kuhn, Thomas S. *The Structure of Scientific Revolutions,* 2d ed. Chicago: University of Chicago Press, 1970.

——— *The Essential Tension.* Chicago: University of Chicago Press, 1977.

Lakatos, Imre. "Falsification and the methodology of scientific research programmes." In Imre Lakatos and Alan Musgrave, *Criticism and the Growth of Knowledge,* pp. 91–195. Cambridge: Cambridge University Press, 1970.

——— *Mathematics, Science and Epistemology.* Cambridge: Cambridge University Press, 1978.

Lakatos, Imre, and Alan Musgrave, eds. *Criticism and the Growth of Knowledge.* Cambridge: Cambridge University Press, 1970.

Lakoff, George, and Mark Johnson. *Metaphors We Live By.* Chicago: University of Chicago Press, 1980.

Laplace, Pierre-Simon de. *Essai sur les probabilités* (1819). English translation, New York: Dover, 1951.

Laskar, Jacques. "Large scale chaos and marginal stability in the solar system." In *XIth International Congress of Mathematical Physics,* ed. D. Iagolnitzer, pp. 75–120. Boston: International Press, 1995.

Latour, Bruno. *Science in Action.* Cambridge, Mass.: Harvard University Press, 1987.

Latour, Bruno, and Steve Woolgar. *Laboratory Life: The Social Construction of Scientific Facts.* Beverly Hills, Calif.: Sage Publications, 1979.

Laudan, Larry. *Beyond Positivism and Relativism: Theory, Method, and Evidence.* Boulder, Colo.: Westview, 1996.

——— *Science and Relativism: Some Key Controversies in the Philosophy of Science.* Chicago: University of Chicago Press, 1990.

Lynch, Michael. *Art and Artifact in Laboratory Science.* London: Routledge and Kegan Paul, 1985.

Maddox, John. "More muddle over the Hubble constant." *Nature,* 376 (27 July 1995), p. 291.

Martin, David W. *Doing Psychology Experiments.* Pacific Grove, Calif.: Brooks Cole, 1991.

Marx, Leo. "Reflections on the neo-romantic critique of science." In *Limits of*

Schmidt, Siegried J., ed. *Der Diskurs des Radikalen Konstruktivismus.* Frankfurt am Main: Suhrkamp, 1987.

Schweber, Silvan S. *QED and the Men Who Made It: Dyson, Feynman, Schwinger, and Tomonaga.* Princeton: Princeton University Press, 1994.

Shapin, Steven. *A Social History of Truth: Civility and Science in Seventeenth Century England.* Chicago: University of Chicago Press, 1994.

Shapin, Steven, and Simon Schaffer. *Leviathan and the Air-Pump: Hobbes, Boyle, and the Experimental Life.* Princeton: Princeton University Press, 1985.

Shimony, Abner. *Search for a Naturalistic World View.* Vol. 1: *Scientific Method and Epistemology;* vol. 2: *Natural Science and Metaphysics.* Cambridge: Cambridge University Press, 1993.

Sinsheimer, Robert L. "The presumptions of science." In *Limits of Scientific Inquiry,* ed. Gerald Holton and Robert S. Morison, pp. 23–36. New York: W. W. Norton, 1979.

Sokal, Alan D. "Transgressing the boundaries: Towards a transformative hermeneutics of quantum gravity." *Social Text,* 46/47 (1996), pp. 217–252.

Sorensen, Roy A. *Thought Experiments.* Oxford: Oxford University Press, 1992.

Stent, Gunther S. "Cognitive limits and the end of science." In *The End of Science?* ed. Richard Q. Elvee, pp. 75–90. Lanham, Md.: University Press of America, 1992.

Strong, J. *Concepts of Classical Optics.* San Francisco, W. H. Freeman, 1958.

Tarski, Alfred. *Logic, Semantics, Meta-Mathematics,* 2d ed. Indianapolis: Hackett Publishing Co., 1983.

Taylor, J. G., ed. *Tributes to Paul Dirac.* Bristol: Adam Hilger, 1987.

Teller, Paul. *An Interpretive Introduction to Quantum Field Theory.* Princeton: Princeton University Press, 1995.

Traweek, Sharon. *Beamtimes and Lifetimes: The World of High Energy Physicists.* Cambridge, Mass.: Harvard University Press, 1988.

Wallace, Philip R. *Paradox Lost: Images of the Quantum.* New York: Springer-Verlag, 1996.

Weinberg, Steven. *Dreams of a Final Theory.* New York: Pantheon, 1992.

——— *The First Three Minutes: A Modern View of the Origin of the Universe.* New York: Basic Books, 1977.

Weiner, Jonathan. *The Beak of the Finch: The Story of Evolution in Our Time.* New York: A. A. Knopf, 1994.

Whittaker, Edmund. *A History of the Theories of Aether and Electricity: The Classical Theories.* London: Thomas Nelson and Sons Ltd., 1951.

Wigner, Eugene. "The unreasonable effectiveness of mathematics in the natural sciences." *Communications in Pure and Applied Mathematics,* 13, no.

1 (1960). Reprinted in *Symmetries and Reflections*, pp. 222–237. Bloomington: Indiana University Press, 1967.

Winch, Peter. "Understanding a primitive society." *American Philosophical Quarterly*, 1 (1964), pp. 307–324.

Wolpert, Lewis. *The Unnatural Nature of Science.* Cambridge, Mass.: Harvard University Press, 1992.

Ziman, John. *Reliable Knowledge: An Exploration of the Grounds for Belief in Science.* Cambridge: Cambridge University Press, 1978.